D0593033

The Petkau Effect

Nuclear Radiation, People and Trees

The Petkau Effect

Nuclear Radiation, People and Trees

By Ralph Graeub

Introduction by Dr. Ernest J. Sternglass

Four Walls Eight Windows
New York

Published by:

Four Walls Eight Windows
PO Box 548
Village Station
New York, N.Y., 10014

First Edition
First printing February 1992

Revised English edition of Der Petkau-Effekt
(Zytglogge Verlag Bern, 1990)
Translated from the German by Phil Hill.

Library of Congress Cataloging-in-Publication Data:

Graeub, Ralph, 1921—
The Petkau Effect / Ralph Graeub.—1st ed.
p. cm.
ISBN: 0-941423-72-7 (cloth)
1. Ionizing radiation—Health aspects. 2. Ionizing radiation—Environmental aspects. 3. Radioecology. I. Title
RA596.G69 1992

574.19"15—dc20 91-26809
CIP

Printed in the U.S.A.
Text design by Blue Brick.
Graphs by Michel Eckersley.

Table of Contents

Foreword

Almost twenty years ago, when I published the book *The Gentle Killers: Nuclear Power Stations Unmasked,* the nuclear establishment contemptuously branded me a "lone voice in the wilderness." Since then, the wilderness has fortunately become much more populated, thanks to the many concerned scientists and citizens that have joined the battle against the threat of nuclear power all over the world, mobilized first by the Three Mile Island accident in 1979 and then by the Chernobyl disaster in 1986. Today, one survey after another indicates that there has been a complete turnaround in public opinion, so that both in the United States and other countries those opposed to nuclear power have become a significant majority. But the question is: for how long will this continue to be the case?

Within the last few years, a new uncertainty has arisen in the minds of many people concerned about the environment. The problem arises from the burning of coal and oil to generate electricity, which not only leads to acid rain that is threatening our lakes and forests, but also produces carbon dioxide that could result in an increase of the average temperature due to the Greenhouse Effect, leading to flooding of coastal areas and climate changes.

Taking advantage of these concerns, the nuclear industry—desperate to change public attitudes—and governments that want to continue production and testing of nuclear weapons, have launched a crusade in the media to promote a new generation of "inherently safe" nuclear plants that would be "environmentally benign" and free the industrial nations from dependence on imported oil from an unstable Middle East.

None of the above claims are valid, as the environmental and alternative press has made clear. As noted by Michael Mariotte, executive director of the Nuclear Information and Resource Service, writing in the *Multinational Monitor,* "despite claims of would-be developers, the new plants are not inherently safe." Karl Grossman, writing in *Extra* of

May/June 1990, points out that even the Nuclear Regulatory Commission's own Committee on Reactor Safeguards reported:

> Accidents can be postulated that would challenge the defense-in-depth concepts being advanced. . . . The two major accidents that have been experienced in nuclear power, those at TMI-2 [Three Mile Island] and Chernobyl-4, were caused, in large measure, by human error. These were not simple 'operator errors' but instead were caused by deliberate but wrong actions. There seems to be little merit in making claims for the improved safety of new reactor designs if they have not been evaluated against the actual causes of the most important reactor accidents in our experience.

As for solving the greenhouse problem, Michael Philips writing in *Sierra* the same year notes that "Replacing all U.S. fossil-fueled plants with nuclear reactors would reduce global greenhouse-gas emissions by about four percent." John Gofman, writing in *Groundswell* in the fall of 1988, argued that nuclear energy actually makes a "net addition to the Greenhouse Effect from carbon dioxide, when fossil fuels burned in uranium mining and refining, reactor construction and clean-up are taken into account." And, with regard to reducing the dependence of the U.S. on foreign oil, Scott Denman, director of the Safe Energy Communication Council, is cited in the *Extra* article as pointing out that only three percent of U.S. electricity is generated with imported oil, and that small fraction, bought largely from Venezuela, is a by-product of gasoline refining used by automobiles.

The present book, which is a revised and updated fourth edition of *The Petkau Effect,* first published in Switzerland in 1985, provides additional reasons why nuclear energy is not "environmentally benign" so that the public that is continuing to be deceived by nuclear propaganda can make a more informed judgement as to whether it really wants a "new generation" of nuclear plants for both bomb-making and civilian power generation.

The aim of this book is to present the range of health and ecological dangers of fission products released into the air and water. Among the most important of the recent scientific discoveries that has been successfully kept from the public is the Petkau Effect, the discovery that showed low-dose, protracted radiation exposures such as those produced by radioactive fission products, to be hundreds to thousands of times as damaging as the same dose received in a short medical X-ray. And, as the present book tries to show, the biological damage is not confined to humans, but applies to all other forms of life, from fish to birds and mammals, and even to our trees.

There is no longer a cold war to justify the continued operation of hundreds of nuclear plants daily releasing highly toxic radioactivity into our air, our milk and our drinking water, constantly adding to the nuclear wastes that no one knows how to keep out of the environment for thousands of years. Only an immediate shutdown of all reactors can end the threat to our health and that of future generations.

Ralph Graeub, October 1991

Introduction

Few areas of biomedical research have been explored as much as the action of ionizing radiation on animals and human beings since X-rays and radioactivity were discovered at the end of the 1800s. Yet in 1972, more than three quarters of a century and thousands of laboratory and human studies later, a simple, quite unplanned experiment on a biological membrane by a Canadian physician and biophysicist named Abram Petkau completely overturned all conventional ideas on the biological damage produced by extremely low doses of radiation.

Suddenly it became clear that the scientific community had been misled into believing that low dose radiation is harmless. Decades of lack of evidence for detectable adverse effects of low dose diagnostic X-rays on human health, numerous animal studies showing very low effects on the genes of reproductive cells, and the relatively low rate of leukemia and cancer deaths among the survivors of the Hiroshima and Nagasaki nuclear bomb detonations lulled scientists into complacency. Studies of effects on plants and trees revealed little sensitivity to radiation, which had been present on Earth in the form of cosmic rays and background radiation from radioactive elements such as radium and thorium ever since the planet's creation some four billion years ago.

Ralph Graeub has been in the forefront of the effort to warn the European public of the dangers of radiation released from nuclear reactors and the attempts by government agencies to cover them up since 1972, when he wrote his first book *The Gentle Killers: Nuclear Power Stations.* This translation of his new book not only incorporates all the new knowledge of the effects of low levels of fission products on human health gained in the intervening year, it also makes a uniquely valuable contribution by including, for the first time for the English-speaking public, a detailed discussion of the newly discovered devastating role of radioactivity in the production of acid rain, ozone and the worldwide death of our trees.

The fact that high doses of radiation produce serious effects was learned very soon after the discovery of X-rays by Wilhelm Röntgen in 1895 and the discovery of radioactivity by Becquerel the following year. Skin burns that would not heal and eventually turned into skin tumors were reported by the pioneers in the application of X-rays to medical diagnosis and therapy within a few years, and most of the early workers in the field died prematurely of leukemia and cancer.

Genetic effects on future generations due to irradiation of reproductive cells before conception were not discovered until 1927, in the course of studies on fruit-flies by H.J. Muller. However, the dose required to double the number of spontaneous mutations was found to be very high, thousands of times the annual dose from background sources such as is received from radioactivity in the soil and cosmic rays from the sun, typically 100 millirads per year. (One millirad is one thousandth of a rad, a measure of the energy absorbed in a gram of tissue.) And in the half century since Muller's discovery, during which radiation was used to deliberately induce mutations in plants by irradiation of the seeds so as to produce more desirable properties in a few of them, similarly high doses were needed to increase the number of mutated plants.

Thus, there seemed to be little reason for concern that ordinary diagnostic X-rays, which gave doses on the same order of magnitude as background radiation, would have any detectable effects on the newborn, or lead to significant increases in cancer. Cancer induction was believed to be caused by similar effects on the genes as that which created mutations of the reproductive cells.

In fact, the early evidence from the treatment of cancer by radiation, where doses of thousands of rads were directed at tumor cells to kill them, showed a remarkable ability of nearby healthy tissue to survive large doses and heal. Spreading the doses over a series of treatments, covering a few weeks to months, helped to reduce any deleterious effects, and radiation therapy became a very widely-used treatment for many types of cancers, further reducing the concern about low dose exposures.

The successful experience with medical uses of X-rays caused the scientists who developed the atomic bomb during World War II to believe that the principal effect of the bomb explosions would be produced by blast and fire, even though they emitted a short burst of gamma rays and neutrons. There was also some fallout from the drifting radioactive cloud believed to produce very low doses, much less than the annual background dose (except for ground bursts near the point of detonation). There was no apparent reason to expect any detectable effects on the environment or health effects in humans from nuclear bomb testing. And since the military was anxious not to have the possibility of biological damage on areas distant from the detonation of nuclear weapons widely known, all aspects of this subject were classified until 1957, when Congress held hearings on the need for fallout shelters.

There was even the hope that there might not be any health effects at all from very small doses close to those from background radiation. A safe threshold on the order of 50-100 rads, a thousand times the small annual background dose seemed to exist. And one study after another of cancer induction for animals in a laboratory setting showed that extended exposures were much less likely to lead to the development of cancer than single short exposures at the same total dose. Thus, it also appeared to be perfectly safe to go ahead with the construction of large nuclear reactors that would bring clean energy, to replace dirty coal plants blackening the air and filling it with the sulphur that produces acid rain. It was widely believed that nuclear power plants would not release any fission products into the environment.

The first indication that there might be a problem from very low dose radiation came quite unexpectedly, from a study undertaken by a British physician at Oxford University by the name of Alice Stewart. There had been a sharp rise in childhood leukemia in England since World War II, so with the help of the British health authorities she sent out a few hundred questionnaires to women whose children had died of leukemia and to a matched control group. To Stewart's surprise, she found that there was no difference in family history or exposure to benzene and other chemicals for the two groups, but that women who had received two or three diagnostic X-rays during pregnancy had nearly twice as many children who died of leukemia before age 10 as the control group, as described in the *British Medical Journal* in 1958.

At first, no one in the medical community wanted to believe her results, but when a similar study carried out by Dr. Bryan MacMahon at Harvard University confirmed her findings in a 1962 article in the *Journal of the National Cancer Institute,* some public health officials became concerned. It was at this time that I became aware of Stewart's findings, in the course of a study with a group of scientists in Pittsburgh of the likely result of a nuclear war, following the end of a temporary halt in testing with the detonation of a 50 megaton hydrogen bomb by the Soviet Union. I realized that the resultant worldwide fallout from this one explosion would give the equivalent of an abdominal X-ray of about 200-300 millirads to some two billion persons in the northern hemisphere. Thus, if Dr. Stewart's data applied also to fallout exposures, there would be something like a 25% increase in childhood leukemia just from this one bomb detonation alone, a result that I succeeded in getting published in June of 1963 in the journal *Science,* hoping that it would help to end the testing of nuclear weapons.

With the conclusion of a treaty to end testing in at least the atmosphere later that year, the widespread concern about fallout in the milk and diet that had been started by Linus Pauling and Andrei Sakharov in the late 1950s with their prediction that millions would die as a result of the bomb tests subsided. But the continuing build-up of nuclear weapons arsenals and the proposed construction of anti-ballistic missile systems carrying nuclear warheads caused me to undertake a study on whether an increase in leukemia rates due to fallout had actually taken place.

In the course of this study, I found evidence for a rise in leukemia rates in the area of Albany-Troy, New York, following a heavy rainstorm that occurred when a nuclear bomb cloud from the Nevada test-site passed over this region. There was also a sharp rise in spontaneous miscarriages and infant deaths within a year of this fallout episode, a result that was published in April 1969 in *The Bulletin of the Atomic Scientists.* Since the number of infants that die of all causes is ten to twenty times larger than those who die of leukemia, the threat to human health from worldwide fallout was now much larger.

Public health statistics indicated that there was an upward deviation from the steady downward trend of about four percent per year for the U.S. as a whole, which ended in the late 1960s after the end of atmospheric tests by the U.S., the Soviet Union and Great Britain. Furthermore, this abnormal increase was very closely correlated with state-by-state data on the concentration of strontium 90 in the milk during the preceding three to four years for all socioeconomic groups. Poorer non-white infants showed about twice the mortality rate than the white population with its better diet, prenatal care and access to medical treatment. The statistics showed that between 1945 and 1965, there was an excess of some 400,000 infant deaths above normal expectations in the U.S. alone.

Again, there was widespread disbelief in the scientific community that such small amounts of radiation could be a factor in leukemia, infant mortality and underweight births. But in the course of preparing a critique of my finding for *The Bulletin of the Atomic Scientists,* Dr. John Gofman and Dr. Arthur Tamplin at the Livermore Laboratories of the Atomic Energy Commission calculated that there probably was such an effect of past nuclear tests, though on a much smaller scale using the risk for genetic effects as a basis. In U.S. Senate Hearings on the Underground Uses of Nuclear Energy in November 1969, they also pointed out that presently permitted doses of radioactive releases by commercial nuclear plants of 170 millirads to the general population would increase the cancer mortality rate by as much as 32,000 per year.

By 1970, Dr. Stewart had greatly extended her study. The results, published in *The Lancet,* showed a direct increase in risk with the number of X-rays taken, without any evidence for a safe threshold, lending further support to the fallout hypothesis and those concerned about secret nuclear plant releases. Moreover, because a few percent of all women had received X-rays in the first three months of pregnancy, Dr. Stewart discovered that the risk for such early exposures was some ten to fifteen times greater than for radiation exposures just before birth. This meant that a dose roughly equal to that due to yearly background radiation, or as little as 50-100 millirads, was sufficient to double the risk of childhood cancer and leukemia, some 1000 times smaller than could be expected from the study of the Hiroshima-Nagasaki survivors.

By that time, I had found rises in infant mortality and underweight births around the Dresden nuclear plant near Chicago. These paralleled the reported gaseous emission of doses calculated to be comparable to those from fallout. Together with similar effects on infant mortality for seven other nuclear facilities, I published these findings in the Proceedings of a Conference on Pollution and Health held at the Statistical Laboratory of the University of California, Berkeley, in 1971. This paper also reported sharp rises in deaths due to congenital defects in Utah following the onset of Nevada tests in 1952, rises in childhood leukemia in nearby Utah and more distant areas such as Minnesota, sharp increases in non-infectious respiratory disease deaths due to bronchitis, emphysema, and asthma in such different areas as New Mexico, Wyoming, New York and Illinois following the onset of nuclear tests, as well as leukemia rises in Utah and Nassau County, Long Island, New York, all reached by the drifting fallout clouds. For Nassau County leukemia deaths at all ages, yearly fallout dose measurements were available that showed an increase in risk of 0.48% per millirad, comparable to that observed by Dr. Stewart for X-rays in the first trimester of fetal development.

Not only infant mortality had shown an abnormal rise in the 1950s across the U.S.A., but so had adult mortality at all ages. This had already been pointed out earlier by I.M. Mortyama in a 1980 article that appeared in *Public Health Reports* and a 1964 monograph published by the U.S. Center for Health Statistics. But there was no biological mechanism that could relate this to the low doses produced by fallout radiation until after the unexpected discovery of Petkau, published in the March 1972 issue of *Health Physics* under the innocuous title "Effect of Na-22 on a Phospholipid Membrane." He wrote that cell membranes which could withstand radiation doses as large as tens of thousands of rads when exposed to a short burst of X-rays without breaking, ruptured at less than one rad when subjected to low intensity, protracted radiation such as that produced by radioactive chemicals.

This finding was completely contrary to all previous observations of biological damage by radiation such as genetic effects, birth defects and cancer induction in laboratory animals or humans, which had shown little dependence on the rate at which radiation is delivered to tissue. In fact, the results of a number of laboratory studies seemed to indicate that when the dose rate was reduced, there was less permanent damage to the genes than for very high rates, presumably due to highly efficient repair mechanisms of the DNA in the nucleus of cells that carry the genetic information.

As Ralph Graeub describes, a series of subsequent investigations by Petkau and his co-workers showed that the cell membrane damage was due to a completely different biological mechanism than the direct hit on the DNA molecules in the nucleus of cells that had been observed at the high doses, and dose rates of atomic bomb detonation or medical exposures. It turned out that the cell membranes were destroyed as the result of the action of a negatively charged, short-lived form of ordinary oxygen, the so-called O_2^- free-radical, produced by the absorbed radiation from the life-giving oxygen dissolved in the surrounding fluid. This highly toxic form of oxygen diffused to the outside of the membranes, where it initiated a chain reaction that dissolved the membrane in a matter of minutes to hours, causing the cell to leak and die.

It became clear that a single O_2^- molecule was sufficient to destroy an entire cell, so that only a handful needed to be produced per cell-volume at very low dose rates. But at high dose rates, many millions would be formed in the same volume in the lifetime of the molecule. This was a form of "overkill," much like the case of a balloon, where a single dart is enough to destroy it, and throwing millions of darts at it only represents a waste of energy. In fact, the more free-radicals are created in a given volume, the more they tend to run into each other, causing them to become deactivated to harmless ordinary oxygen. Thus, per unit of energy deposited in living tissue consisting of cells, high doses given at the rate of 10,000 rads per minute were found to be 100 billion times less efficient in destroying a cell than at one ten millionths of a rad per minute, the rate at which we experience background radiation.

As described in *The Petkau Effect,* the consequence of the enormously greater efficiency of radiation at low, as compared to high, dose rates is that the dose-response curve rises very rapidly at small doses and dose rates near that given for background radiation, and flattens out at the high doses and high dose rates for environmental exposures. Mathematically, this turns out to be of the form of a logarithmic relation between dose and biological response for the case of individuals exposed to different amounts of radiation during a given time period, as in the case of releases into the environment. By contrast, in the case of individuals exposed to a series of short medical exposures, where the rate at which the radiation is received is the same for every individual exposure, one gets a linear, straight-line relation between the dose and the response that is hundreds of times less steep, so that the low dose of a few millirads produced by a modern chest X-ray in the case of an adult represents a risk that is thousands of times smaller than the same dose from fallout accumulated in the body of an infant or older person over a period of weeks, months or years.

Thus, the Hiroshima-Nagasaki data as reexamined by John Gofman in his 1990 book *Radiation Induced Cancer from Low Dose Exposure*, shows a dose-response that is concave downward, obeying a fractional power-law or a logarithmic form for the risk of cancer that rises rapidly at low doses and more slowly at high doses. The greatest doses were received at a higher rate in rads per minute than the lower doses of the more distant survivors. But since the more distant survivors also received fallout from the "black rain" that entered the drinking water and diet and thereby produced an additional dose on the order of tens of rads which was not taken into account when the data were analyzed, the very steep initial part of the dose-response curve was hidden and instead had a nearly flat shape. Evidently, in the attempt to deduce the cancer risk near background levels by curve-fitting, extrapolation from high doses leads to an underestimate of the true risk of small protracted doses by a large factor, more so for a linear than for the fractional power law used by Gofman.

As discussed by Graeub, the indirect, free-radical type of damage that dominates at low dose-rates is particularly serious for the cells of the immune system, which must be constantly renewed from their progenitors in the bone marrow. This is especially true for strontium 90 and other bone-seeking isotopes chemically similar to calcium that concentrate in bone and emit relatively long-range beta particles or electrons. These emissions reach the marrow much more efficiently than the alpha particles emitted by naturally occurring radium in the environment. Laboratory studies of animals given radioactive strontium by Stokke and his co-workers at Oslo University, published in 1968, have not only shown that the damage to marrow cells rises rapidly for doses in the millirad range and then levels off to a plateau, but as found by Heller and Wigzell at Upsala University in 1977, radioactive strontium acts to inhibit the normal function of the so-called Natural Killer (NK) cells that originate in the bone marrow and are vital in the defense against micro-organisms and cancer cells.

An equally serious danger is the premature birth of infants whose mother's immune system has been damaged, leading to the rejection of the fetus as a foreign object. This biological mechanism explains the delayed action of strontium 90 in the milk, as it builds up in the mother's body in the years before pregnancy, another unanticipated phenomenon only recently recognized. The small ruptures of blood vessels in the brain of babies born below normal weight greatly increase the chance of neurological damage, in addition to the effect of radioactive iodine slowing the normal development of the brain. As the author describes in detail, the result is an increased risk of learning disabilities as manifested by a decline in Scholastic Aptitude Test (SAT) scores, first reported by Steven Bell and myself at a meeting of the American Psychological Association in September 1979. It also leads to a greater number of behavioral problems, such as an increased incidence of criminal violence, as discussed by Robert J. Pellegrini in *The International Journal of Biosocial Research* in 1987.

As Ralph Graeub makes clear, it is therefore the subtle effects on the hormonal and immune system by the indirect chemical action of manmade radioactivity in the human body which explain the unexpectedly large effects on prematurity, infant mortality, infectious diseases and cancer.

The increase in mortality for all age groups demonstrating the much greater risk of biological damage from the Petkau Effect has been confirmed in a series of very carefully planned and executed epidemiological studies. They involve individuals exposed to low-level radiation in the environment produced by bomb tests, by commercial nuclear plant releases, and workers in government nuclear facilities, as summarized in the 1990 BEIR V report by the U.S. National Academy of Sciences. This report also confirms the earlier evidence for the serious effects on intelligence and school performance of children exposed to radiation *in utero* or early childhood described in the present book.

More recently, a large epidemiological study of 1,117 individuals living in Utah, with known exposure to Nevada test fallout who died of leukemia was published by Stevens and associates in the August 1, 1990 issue of *The Journal of the American Medical Association.* The study showed a clear association between the radiation doses at levels comparable to those from the natural background dose and leukemia mortality, which was strongest for those under the age of 19, consistent with five previous studies that had been questioned for various methodological reasons overcome by the latest study. And in the March 20, 1991 issue of the same journal, an even more extensive study by Wing and co-workers involving 8,318 men individually monitored for decades at the government's Oak Ridge National Laboratory showed that the risk of dying of leukemia, cancer and other causes of death was ten times higher than expected on the basis of the Hiroshima-Nagasaki survivors.

Tragically, as this book makes clear, the discoveries of Stewart and Petkau and their widespread epidemiological confirmation came long after nuclear weapons were tested on a large scale, and long after hundreds of military and civilian reactors were built all over the world. Moreover, in an effort to minimize the danger of accidental releases of radioactivity to difficult-to-evacuate urban populations, the reactors were often located in rural dairy farming areas that supply most of the milk for the large cities by refrigerated tank trucks, a highly efficient pathway for the shortlived radioactivity not considered by the regulatory agency in the original estimates of the total dose to the population. Compounding the tragedy was the fact that in the effort to help poor mothers achieve a better diet during pregnancy, contaminated surplus milk and cheese that often comes from the areas around the large reactors was unwittingly distributed to poor pregnant women by welfare agencies.

As Ralph Graeub concludes, the most difficult task for society and the scientific community today is to face totally unexpected results, and to end all releases of fission products into the environment, even though hundreds of costly reactors that are already operating around the globe must be phased out.

But a failure to recognize our past mistakes can have only the most tragic consequences, in continued high rates of underweight births, as well as rising mortality due to new types of infections such as AIDS, or the resurgence of old ones like tuberculosis. And the same is true for cancer mortality, as the ability to detect and destroy cancer cells is increasingly impaired with the build-up of strontium in the bone of young and old.

Such an unexplained sharp increase in cancer rates has recently been reported by Devra Lee Davis and her colleagues in a 1990 volume of the *Annals of the New York Academy of Sciences,* dedicated to the problem of rising cancer mortality during the last two decades, especially among individuals over the age of 55 in England, what was then West Germany and the U.S.A. These are exactly the nations where fission products have been released into the environment from nuclear reactors in the greatest amounts since the end of the 1960s.

But the greatest danger lies in the future of the large cities, which have been the centers of civilization since its beginnings. The reason is that the immune system organs of the newborn infants continue to be damaged by radioactive strontium imported with the milk, the cheese, the vegetables, the fruit and the meat produced in rural areas near the large reactors. The same is true for the function of the fetal and infant thyroid, controlling growth and development by the shortlived iodine in fresh milk, affecting all social classes, but especially the poor. The result is a continuing tide of underweight births that not only leads to a high rate of birth defects, infectious diseases and cancer, but also produces an impaired learning ability. This increases the chance of poor school performance, which reduces job opportunities in a high-tech society.

At the same time, young females who drop out of school are more likely to become pregnant in early adolescence and give birth to more underweight babies supported by welfare in a single parent family, thus perpetuating a vicious cycle of poverty and hopelessness that drives up welfare and health-care costs and forces cities into mounting debts or insolvency. Employment, income and tax revenues decrease, requiring a further reduction in social services and educational opportunities for the poor, thereby accelerating the tragic process of decline.

Thus, as is clear from the evidence presented in *The Petkau Effect*, that in a technological world where learning ability is the key to productivity, competitiveness and a high standard of living, it is no longer possible to refuse to face the role of low-level radiation in aggravating our existing social problems. The enormous effect of minute doses explained by the Petkau Effect can only accelerate the disintegration of our cities as the middle-class flee to the suburbs, and the spiral of disease, underweight births, single parent families, poverty, hopelessness, drugs, and crime destroys the fabric of society.

The danger is not only to humans and animals. Ralph Graeub presents enormously important but little-known evidence that nuclear plant releases are also contributing to the production of acid rain, ground-level ozone and the death of the forests. As Graeub points out, the death of trees has reached epidemic proportions not only in Europe, but also in many parts of the United States downwind from both ordinary and nuclear power plants. Again, the indirect effects of radiation through the production of free-radicals on ordinary air-pollutants and the cell membranes of plants seem to be the reason why the effects of low doses are so much more damaging than had been expected on the basis of high dose studies.

But emerging from all the disturbing new knowledge that this book contains, there arises the hope that the threat to the health of human beings, animals and trees can be ended. This has already begun to happen in many parts of the world where fallout levels from bomb testing have declined, and no new radioactivity has been added to the environment by "the gentle killers," the peaceful nuclear reactors that the scientific community had hoped would atone for the horrors of the atomic bomb.

Ernest J. Sternglass
Department of Radiology
University of Pittsburgh
School of Medicine

I. ECOLOGICAL CONSIDERATIONS

Ecosystems—Havens of Life

With the official approval of all nations, and under the eyes of the experts, humankind has slowly worked itself into a serious predicament. Blind overestimation of material progress and its comforts are the immediate reasons for this development. Deeper causes, however, are to be found elsewhere: for the most part of the past, decisive century, humankind has completely ignored ecological laws, or at least not taken them seriously.

Classical biology had for too long been concerned only with individual creatures. It was, after all, doing so in accordance with an ethic which has now been called into question. This ethic is built around the well-being, the dignity and the survival of the individual human being, and demands that these be maintained under all circumstances. In botany and zoology, this way of thinking expressed a general tendency to identify only individual plants, or study the anatomy of individual animals.

More recently, however, there has been a realization that according to a wider view of "life" (which necessary means the totality of life on our planet), it is unacceptable to accord human beings the central position. Rather, it is necessary to recognize and support the interrelationship of the living world, of living communities in their environment—i.e., life in its ecosystems. This new science is called ecology. Its cautious beginnings can be traced back to 1864.

Ecology demonstrates that nature has united "life" on earth—i.e., plants, animals *and* humans—into a natural balance, a natural economy. In the security of this natural balance, life should be able to continue for millions of years—in effect, indefinitely. The natural balance of the ecosystems involves no increase in environmental pollution and no irretrievable waste products. Instead, wastes are rapidly made available for the formation of new life. Consider for a moment how this natural economy functions. Under the laws of ecology, the human species is an ordinary organism that integrates in the same way as plants and animals, in spite of its intellect, its intelligence, its language—in short, in spite of its supposed special place.

Five Organizational Levels of Ecosystems

Life operates at five different levels which are linked together and with their environment by a network of feedback mechanisms.[192] The levels are:

1. *Individual cells*, such as bacteria and algae. They are independently viable, and represent the most primitive stage of life.
2. *Multicellular organisms*. Every animal, plant and human being falls under this category. Here, the cells have abandoned their independence; they are specialized and highly interdependent. This is the first stage in which we find nervous systems.
3. *Populations* are reproductive communities of organisms of the same species. Humankind as a whole represents this population, as do all animals or plants of a given species. The chief characteristic of a population is to serve as a carrier and transmitter of genetic inheritance.
4. *Ecosystems* Yet not even populations are independent units. In nature, all plant and animal populations in any area live in extensive dependence on one another and their habitat. These living communities form ecosystems. Every human settlement must be considered to form this system, together with its habitat.

   At the ecosystem level, physical and chemical cycles, including energy cycles, take effect. In this way, an ecosystem is comparable to an organism in which each component has a task which it performs. Therefore, any injurious interferences with this organism can have serious consequences.
5. *The Biosphere*. This constitutes the fifth and last stage. It is the sum of all the ecosystems on our planet, which are, once again, interrelating with one another. The biosphere is an all-embracing organism of which we humans are only a small part.

The Ecosystems' Three Levels of Production

Moreover, the natural economy of the ecosystems, like human society's version of mercantile economy, has producers and consumers. A comparison of the two economic systems provides the key to an understanding of ecology.

1. *Producers.* In nature, the plants have this role. Using solar energy (photosynthesis) they produce trees, grass, weeds, fruits, vegetables, etc.—in short, they miraculously produce themselves from the minerals in the soil, and from the carbon dioxide in the air and water. They give off oxygen as a waste product.
2. *Consumers.* Plants and their fruits then serve as food for all the other members of the ecosystems—the consumers, animals and humans. Even exclusive carnivores feed on prey which ultimately live off plants.
3. *Decomposers.* In the natural economy, there is a third group, the decomposers. They eat or use the waste of the producers and consumers. Decomposers include, for example, fungi, bacteria, rain worms and many small organisms in the humus. With oxygen, they transform dead plants, fallen leaves, and also the feces and corpses of consumers, into minerals, carbon dioxide and humus. The oxygen becomes a waste product of the plants. Thus, waste is rapidly put to the service of the producers—i.e. plants—in their construction of new life. The cycle is then complete—a natural recycling process. This living community—this ecosystem—can continue for an unlimited period, for in addition, its energy input occurs by means of renewable energy carriers: pollution-free solar energy and food materials. Thus, nothing is lost in a natural economy.

By contrast, the ecosystems of the superimposed "civilized" economy operate almost completely linearly, from the producer to the consumer and then straight to the waste dump. In other words, the raw materials are lost for all time. This type of an economy can exist for only a short time. Many waste materials (pollutants of all types) generated both by producers and by consumers are habitually fed at random into the natural economy ("disposed of"), where they damage or destroy the complicated networks of life. The death of forests, generally attributed to acid rain, is the latest example. The deterioration of the foundations of life too, is accelerated by the systematic expansion of civilized economy (economic growth).

Regulation Mechanisms of Ecosystems

Nature is only peaceful in appearance. In reality, a stern law of eat and be eaten is in force. It is the precondition for a healthy ecosystem. Natural selection causes everything that is weak or sickly to be eliminated, and (very rare) favorable mutations (alternations of genetic characteristics) enable higher development (evolution) to take place.

The question "who eats whom?" cannot, however, be answered by referring simply to a food chain such as: grass—locust—frog—snake—bird of prey. Rather, a complex law of feeding is in force. Complicated regulation mechanisms ensure a dynamic equilibrium. Neither producers nor consumers and decomposers can outgrow certain population limits. Otherwise, the entire ecosystem would collapse.

As stated above, the fact that animals not only eat, but are also eaten, constitutes an important regulatory mechanism. Herbivores are controlled by carnivores. Thus, when humans crowded out foxes, wolves and pumas, the natural enemies of deer and rabbits, the latter increased in number to a disastrous degree. Carnivores too, are kept in check by other carnivores. All young animals are constantly threatened by raiders and accidents. Moreover, a predator that feeds on various prey will most frequently hunt the more common ones, so that automatic limits are placed on population increase. In addition, even the strongest carnivore can fall victim to parasitic diseases, the greater the number of its species found in the same area.

In this way, "life" in an ecosystem is guaranteed long-term security and stability, albeit at a high cost to individual life. The biological unit of the ecosystem is thus comparable to an individual organism in which individual cells are continually dying off and being replaced, controlled by complex nervous and glandular regulators.

True, even undisturbed ecosystems are not stationary in the long term—but this, too, is "upward mobility." In the mutual interaction between natural selection and mutation, an ecosystem, like a living organism, learns how to improve energy utilization through the cycles of trial and error. Short-lived organisms are increasingly replaced by long-lived ones. In this way, ever more complex communities with ever more highly developed organisms have been formed over the course of our planet's history. We have to recognize that new creations have been formed in this process which are beyond our comprehension. The huge gaps in scientific knowledge are a fact of life.

Tropical Forests as an Example of an Ecosystem

Tropical rainforests, such as may still be found in the central African jungle and the Brazilian Amazon basin, are fine examples of ecosystems with closed food chains. These rain forests constitute closed, self-sustaining systems, with plants and animals as producers, many animals as consumers, and termites, worms and bacteria as decomposers. These subsystems live separated from one another in various horizontal levels between 1.5 and 35 meters [four and 100 feet] above ground level, where hardly a ray of sunlight falls. In these impenetrable, yet intertwined, forests there is no change of seasons. Death and regeneration flow into one another unceasingly. At least 70% of all nutritive substances are contained not in the soil but in the living realm above.

However, the green magnificence of the tropical rainforest conceals the fact that its soil is infertile. To cut down the rainforests is to gain farmland for only a few harvests; thereafter, an infertile desert remains.

Up until thirty years ago, there were still 16 million square kilometers of rainforest. Eight million remain today, and the clear-cutting continues relentlessly. The continuing extinction of these forests will affect world climate in an incalculable manner. 5

Comparisons: Natural vs. "Civilized" Economies

Approximately three decades ago, there was no discussion of ecological relationships. Today, we know that ecosystems remain largely intact only when the civilized economy superimposes itself on the natural economies of our ecosystems in an appropriate manner.

For this reason, four important points should be presupposed:

- the cycle principle
- the concept of "zero emissions"
- renewable energy carriers
- constant populations of producers and consumers, and also of "decomposers."

The Cycle Principle

Our economy lacks decomposers for its industrial waste products and for the pollutants emitted during production; in addition, the natural cycle is interrupted. As we continually harvest our fields, which can only be kept fertile with artificial fertilizer, we are not able to return wastes to them. Urbanization has contributed to this problem, but so has the constant production of toxic and non-biodegradable materials. Lack of humus, erosion, attacks by pests, over-fertilization, and also poisoned soil and unhealthy foods are the consequences of such misdirected agriculture. Organic farming is a way out of this, but only if we stop filling our environment with pollutants. These include dangerous artificial radioactivity from nuclear power.

The Concept of "Zero Emissions"

As forests are destroyed, we will have to envision a civilized economy based on the "no-threshold" principle, which would mean aiming for zero emissions.[37] Of course, this cannot be accomplished overnight, nor even by next week. For not only purely technological problems exist, but far-reaching socio-political ones as well.[37]

The build-up of heavy metals (such as PCB, DDT and pesticides), compounds which are very difficult to biodegrade, and artificial radioactivity in the biosphere are warning signs. We will have to curtail doing and producing things that release toxic emissions into our air.

Renewable Energy Carriers

Our energy carriers are for the most part non-renewable. We are devouring our limited supply of fossil fuels, which we have until now used much too carelessly. In addition, nuclear energy is destructive to all life forms, and therefore constitutes an ecologically unacceptable complement or alternative. Increasingly, it is contaminating our habitat; in fact, our entire biosphere. Only solar, wind and water-based energy, as well as (possibly) geothermal energy are inexhaustible, clean energy carriers. Ultimately, we will have to orient ourselves toward them.

We don't have much time, and since an immediate restructuring is not possible, fossil-fuel-based energy production (of petroleum, and coal) will have to be accomplished in a more environmentally benign manner. This can be done rapidly and economically. That will mean, however, giving an equally high priority to the dismantling of our nuclear power system, since its production facilities cannot possibly achieve any significant reduction in emissions.

Moreover, ecosystems always strive for a condition of minimum necessary energy flow. The civilized economy, too, will in the future have to achieve much better utilization of its energy. Great reserves can be tapped to avoid loss and yield higher degrees of efficiency. Energy savings need not mean reduced comfort.

Reduction of atmospheric pollutants to the desired level of the 1950s would mean an absolute end to any increase in radioactivity. After all, the build-up of these pollutants in the ecosystem did not really start until that time. Radioactivity, more than any other pollutant, can have multifaceted synergistic—i.e., damage-amplifying—effects.

For this reason, research appropriations for nuclear technology must be eliminated, and existing projects must be terminated. Only when research funds are committed to renewable energy, can these resources be implemented technologically and economically.

Constant Populations of Producers and Consumers 7

Humankind has gradually removed itself from the effect of nature's regulatory mechanisms—in particular that of natural selection. Our weapons, science and technology, became and still are medical advances used to vanquish all enemies. However, our misguided ethic to strive first and foremost for the well-being, dignity and survival of the individual human being regardless of the overall risk to the human population, has been just as decisive. One consequence is the paucity of sensible family planning.

We certainly know that humankind is increasing explosively outside the industrial nations. According to the 1984 UN conference in Mexico City, world population will increase from 4.7 billion people to 6.3 billion by the year 2000. Thus, within the next decade, the population will have increased by 1.5 billion. This predicted boom is considerably more than the current population of Europe. In order to create dignified living standards, an infrastructural system of continental proportions will have to be maintained, and within a short span of time. Aside from its political and economic consequences, there is neither the time nor the resources to achieve such a task.

Of course, humans do not want to fall back into a primal state in which natural selection operates. For this reason, a stabilization of the population is imperative. We have a duty, too, toward the helpless animal and plant kingdoms, for we alone have the gift of prescience, of reason and also of a compelling conscience.

However, stabilization of the population must be approached with great discretion for chemical and radioactive pollution are so great a threat that the birth of a healthy child may soon be a precious event.

There is another important aspect to consider: long ago, the sick and the weak were removed from among the healthy population by plagues, famines, and other diseases and disasters. This gruesome selection reduced the reproduction of genetic defects considerably. Thanks to medical advances, modern society is able to often keep the weak and sick alive. This leads humankind to a new interpretation of responsibility. Previously, thanks to natural selection, although natural radioactivity was always encountered by humans, its damage was fleeting and not passed on to future generations. Now there is no longer any question that, whether artificial or natural, genetically damaging effects start at the zero-dose point of radioactivity. The assertion that we have since time immemorial lived unharmed in a bath of natural radiation is fundamentally incorrect, since it ignores the original process of survival. Natural radioactivity, too, has always been too much radioactivity, and any additional increase in the level of radiation must be prevented. As we all want to protect the sick, the weak and the genetically damaged, we will also have to accept the responsibility of caring for our greatest and most irreplaceable possession, our pool of genetic information. It is irresponsible to emit ("dispose of") proven mutagenic (genetically damaging) substances into our habitat, as is ruthlessly and inevitably being done by nuclear technology. These fission products can eventúally, as we shall see below, accumulate in the biosphere and in organisms.

Summary

Particularly during the past century, humankind has acted irresponsibly toward ecological nature. It is as if, at random, pieces or organs (like the kidney or the liver) have been cut out of an organism, without consideration of the resulting consequences.

The more multifaceted and complex an ecosystem becomes, the greater the stability it demonstrates against outside interference. With every inappropriate civilized intervention (extinction of animal species, intervention in chemical and physical cycles through all types of toxic pollution, interference with the genetic inheritance of plants and animals, clear-cutting of forests, regulation of riverbeds, etc.) there are unforeseeable consequences. Unfortunately, humankind has already severely damaged its habitat, and thus the basis of its own existence. Forest death has indeed become a new sign of the threatening ecocatastrophe.

The only way out of the existing deplorable situation is through an overarching ecological evaluation of all life-issues. Unlike other sciences, ecology cannot "boast" of being value neutral. On the contrary, an ecology without ethics and morals, without emotions and any responsibility toward the future, is inconceivable. Anyone today who feels no empathy when considering the future of a starving people or of the endangered animal and plant kingdoms has probably already lost a portion of his or her humanity.

II. ATOMIC BOMBS AND ATOMIC POWER (Biological Effects)

Foundations of Nuclear Physics

Structure of the Atom

In considering the dangers of nuclear power plants, it is useful to discuss some of the fundamentals of nuclear physics. The following discussion of the atom's structure is designed to be generally comprehensible, since it is not necessary to have a thorough understanding of a subject in order to be able to form one's own judgment—although the exact opposite is always being claimed.

In perspective, an atom can be compared with our solar system. In the center is the sun, circled by the planets. Analogous, but on an unimaginably small scale, the atom consists of a *nucleus* (corresponding to the sun) surrounded by *electrons* (corresponding to the planets).

The mass of the atom is almost entirely concentrated in its nucleus. Thus, for instance, the nuclear mass of a hydrogen atom is some 2000 times as great as that of an electron circling a nucleus. In addition, atoms are virtually "empty." If one imagines the nucleus to be the size of a hazelnut (approximately 1 cubic centimeter), the electron would orbit a nucleus at a distance of about 500 meters, or one third of a mile! The nucleus is incredibly dense. A cubic centimeter of pure nuclear mass will weigh 240 million tons.

Ionization

The nucleus consists of protons with a positive electrical charge, and neutral particles, or neutrons. Normally, the number of protons in the nucleus corresponds to the number of negative electrons, so that the atom is electrically neutral. If, however, there are more or less electrons present in the electron shell than there are protons in the nucleus itself, the atom will have either a positive or a negative charge, i.e., will be ionized, and constitute a so-called ion.

Ionizing Radiation

Radiation having the capacity to knock electrons out of the electron shell is known as ionizing radiation. In such a case, the atom is positively charged, or ionized, due to the loss of negative particles—its electrons. A living organism requires no ionizing radiation in order to maintain life. On the contrary, any exposure to such radiation is extremely hazardous, for ionization detrimentally affects, breaks or destroys chemical bonds. Cell toxins may form in the body, which interfere with metabolism and hormonal action, or cause any number of diseases such as leukemia, cancer, etc., as well as genetic defects.

Isotopes

The number of protons in the atomic nucleus is exactly set for each of the elements (over one hundred). These range from the lightest element, hydrogen, which has one proton in its nucleus, to the heaviest elements, one of which is uranium with 92 protons. By contrast, the number of neutrons in the nucleus can vary for those atoms of the same element. These sister atoms are called isotopes.

To identify the elements and isotopes, the mass number (the number of particles in the nucleus) is written behind the element designation. Thus, for instance, uranium 238 has 238 particles in its nucleus, while its isotope uranium 235 has only 235 particles. Both, nonetheless, are uranium.

Radioactivity

There have always been atoms whose atomic cores lacked stability, but which disintegrated without any external intervention. These atoms are known as radionuclides, and their disintegration is called radioactivity. Thus, in addition to the normal carbon 12, there is radioactive carbon 14—a radionuclide. The radioactive disintegration of the unstable nucleus takes place by emission of radiation (ionized radiation). This disintegration progresses according to precise physical laws until a stable nucleus has been formed. Thus radioactive carbon 14 is transformed into nitrogen, tritium (a special name for hydrogen 3) into helium, and phosphorus 32 into sulphur. If they are incorporated into the body, such radioactive substances can cause very serious biological disruptions.

The term natural radioactivity refers to radioactivity occurring in nature without human involvement. Artificial radiation stems from radionuclides generated by human manipulation, such as atomic energy. It is possible to artificially transform stable atomic nuclei into radioactive ones by means of nuclear fission.

The discovery of radioactivity and of ionizing radiation by the German physicist W. C. Röntgen (1845–1923), did not occur until the end of the last century. The X-rays used in medicine still bear his name in German (Röntgenstrahlen); they are generated by electron tubes. Analogous ionizing radiation can occur by other means as well, including during the disintegration of radionuclides; this is called gamma radiation (gamma-rays). Like X-rays, these are electromagnetic waves. Thus, the nucleus emits not a particle, but pure energy. The amount contained in a gamma-ray is considerably greater than that in an X-ray, and consequently has much greater penetrating power. Nuclear reactors require concrete walls of two or three meters in thickness in order to achieve adequate protection against gamma-radiation.

Research into radioactivity began in 1896. Two years later, Marie and Pierre Curie discovered radium and polonium in uranium pitchblendes. In 1910, for the first time, Marie Curie succeeded in isolating 0.1 grams of radium. She did not, however, realize that the mysterious glow of this substance is connected with highly dangerous radiation. She died of leukemia in 1934, one of the first victims of the radioactivity which humankind had now begun to manipulate.

Half-Life

Every radioactive element requires a precise time for its disintegration. The speed of this disintegration is called its half-life, which indicates the time period after which half the nuclei of a definite radionuclide will have disintegrated. This period ranges from fractions of a second to billions of years. The table below shows the half-lives of several radionuclides emitted into our habitat by nuclear facilities:[1]

Isotope:	*Half-Life:*
Strontium 89	50.5 days
Strontium 90	28.5 years
Ruthenium 106	368 days
Iodine 129	15.7 million years
Iodine 131	8.04 days
Cesium 134	2.06 years
Cesium 137	30.1 years
Plutonium 239	24,390 years
Xenon 133	5.29 days
Krypton 85	10.76 years
Tritium (Hydrogen 3)	12.3 years
Carbon 14	5,736 years

Thus, in the case of tritium (radioactive hydrogen), with a half-life of 12.3 years, we find that half of a kilogram of tritium, or about a pound, will still be in existence after 12.3 years.

In the case of plutonium 239, which is produced exclusively by nuclear fission, it takes 24,390 years to reduce it to half its original quantity. Therefore, we are producing wastes which in effect will be with us forever. Just one millionth of a gram of plutonium will cause long-cancer—and tons of it are being produced in nuclear power plants. Yet in newspaper ads published in 1984, Swiss pro-nuclear propaganda claimed that future generations would probably be grateful to us for the nuclear wastes we're leaving them.

Even short-lived radionuclides can be very dangerous, depending on the retention capacities of the nuclear facilities. This is especially true of isotopes of the so-called "rare-earth" elements. They constitute 60% of all radioactive fission-products. Their path through the food chain to the nearby population is a short one. The effect of these elements is very intensive, because all their energy must be expended within a very short time. Iodine 131 should also be mentioned in this connection; it builds up in the thyroid gland.

The inert gas krypton 85 (half-life: 10.7 years) is being emitted in vast quantities, especially by the processing centers. It could have serious consequences for atmospheric electric conditions, and hence for the weather and possibly plants.[34]

Types of Radiation

Four types of radiation are emitted by disintegrating atoms:

1. *Alpha Rays.* These are particle rays of high velocity emitted by the disintegrating nucleus. These particles are relatively "large"—they are nuclei of the helium atom. However, alpha rays only travel for a few centimeters in the air, and only 0.1 millimeters into organic tissue. On the other hand, they ionize powerfully and densely. If these rays manage to penetrate a cell nucleus, they have the effect of a bull in a china-shop—they destroy the structure.
2. *Beta Rays* consist of electrons which also originate in the atomic nucleus. A neutron too, can disintegrate into a proton and an electron, in which case the electron is radiated out. Beta rays radiation can penetrate through several centimeters of organic tissue.
3. *Gamma Rays.* As has already been explained, this is not a particle but rather a highly energetic electromagnetic wave which, unlike alpha and beta rays, can penetrate even concrete, lead and steel. They are almost always generated at the same time as alpha and beta rays.
4. *Neutron Rays.* Neutrons are the electrically uncharged "building-blocks" of the atomic nucleus; they consist of the union of one positron and one electron. They are principally emitted during transformations of the nucleus, as in the explosion of an atomic bomb or the fission of the nucleus in an atomic power plant. Neutrons have powerful, penetrating force; even lead provides only a poor barrier to them, although large quantities of water or paraffin are better.

In summary: it is unnecessary to memorize this data. It is enough to understand that extremely dangerous artificial radioactivity and radiation (of four different types) which are destructive of life are generated. Nuclear power is guilty of the increasing contamination of our entire habitat with these artificial radionuclides. Nuclear power plants are not radiation-tight.

Measuring Radiation

Now that the four types of radioactive radiation have been explained, we should familiarize ourselves with the yardsticks used to measure its physical and biological effect.

1. *Activity:* curies. Activity, or quantity of radiation, is measured in curies. If 37 billion atoms of a substance disintegrate every second, there is one curie of activity; this is the case for even one gram of radium. Since the unit 1 curie is much too great for natural processes, fractions of a curie are often used in calculations:

1/1.000 curie (thousandth) Ci	=	1 millicurie (mCi)
1/1.000.000 (millionth) Ci	=	1 microcurie (uCi)
1/1.000.000.000 (billionth) Ci	=	1 nanocurie (nCi)
1/1.000.000.000.000 (trillionth) Ci	=	picocurie (pCi)

Measurement of the quantity of radiation emitted by nuclear power plants is often indicated in curie units. However, they can easily have a trivializing effect and be misinterpreted, since a low number of curies does not necessarily mean a low level of hazard. A radioactive material's level of danger depends essentially on the type of radiation it emits, the amount of energy involved, the half-life of the substance, and its ecological and biological behavior.

2. *Pure radiation dose:* This dose refers to the quantity of energy absorbed from radiation by one kilogram of a body (either living or inanimate). The purely physical unit for this radiation dose is the rad (for "radiation-absorbed dose"), which is calculated as follows:

$$\frac{1 \text{ Joule}}{\text{kg}} = 100 \text{ rads}$$

One thousandth of a rad is called a millirad (mrad).

3. *Biologically-effective radiation dose:* In addition, the varying biological potential of radiation on living body weight has to be taken into account. Alpha-rays have a volley-like effect at short distances; beta and gamma rays penetrate more deeply. Thus, the purely physical effect cannot necessarily be equated with the biological effect. For this reason, an additional dose unit, the rem (for "radiation-equivalent men"), has been introduced. One thousandth of a rem is called a millirem (mrem).

The fact that the same physical effect (pure energy absorption, measured in rads) can have a varying biological effect (effective dose, measured in rems) can be seen from the following comparison cited by Bodo Manstein: If a living body is on one occasion struck with a pointed object, and on another occasion with a blunt object, each time with the same force (energy), very different wounds will result. Thus, the same physical force has shown differing biological effects.

The various types of radiation, too, can have differing biological effects (rem) with the same physical dose (rads). This is what is meant by the term "relative biological effect," which has come to be expressed by a "quality factor" (QF). Thus, rads can be converted to rems by means of a simple calculation using the quality factor:

$$\text{rem} = \text{rad} \times \text{QF}.$$

For example, a QF of 20 has been assigned to alpha rays, while neutrons get a QF of 10.[84] Thus, these rays are estimated at 20 and 10 times as biologically effective, respectively, as gamma rays:

1 rad of alpha rays	=	20 rem (QF = 20)
1 rad of beta rays	=	1 rem (QF = 1)
1 rad of gamma rays	=	1 rem (QF = 1)
1 rad of neutrons	=	10 rem (QF = 10)

The Rem Lie

The rem is highly controversial. Data in rem give the appearance of accuracy which in fact does not exist. According to Manstein, the qualitative description associated with rem is based on rough estimates which can in no sense take complex biological processes into account. It is impossible to pack both the type and energy of radiation, and also the chemical conditions and changes into one and the same concept. Consider, too, that supposedly identical tissue (e.g., lung, liver, glands) may include millions of cells of the most variegated structures, with numerous functions and sensitivities.[115]

Moreover, direct measurement of the doses of radionuclides incorporated into the body is not even possible; complex calculations and measurements have to be undertaken. If there is a major lack of uniformity in absorbed doses (as is usually the case), which can especially involve the lungs and the bones, the doses become even more unrealistic. The International Commission on Radiological Protection states in this connection:

> There are certain conditions of radiation exposure in protection work where the QF concept as outlined above can only be applied with major qualifications.[69]

UNSCEAR goes so far as to say:

> Models to relate the temporal distribution of absorbed doses from a radionuclide to that of fractionated external irradiation on the basis of equal effects have not yet been fully explored. There is also uncertainty concerning the microdistribution of radionuclide energy. . . , and they affect the assignment of precise values of relative biological effectiveness. . . [205]

Rausch also points to impressive non-homogeneity of radioactive materials distribution in organs. "In bones, for instance, it appears that certain radionuclides occur principally in the periosteum during certain phases of absorption, while others appear in certain sleeve-like microstructures of the compact bone. In the liver, gall bladder and bone-marrow, the deposit of certain radionuclides may occur in the form of so-called 'hot spots,' while major areas of the organ remain free of them. The same is true of plutonium in the bronchia, the lungs and the lymph nodes of the lungs after inhalation, except that here the spatial non-homogeneity has an additional dynamic component in the transportation process of the movement of absorbing cells."[140]

This shows, Rausch says, that the biological effect of major types of radiation can be ascertained only very imprecisely. However, exact data should constitute the most basic precondition for any statement on possible damage, risk calculations and, hence, establishment of limit values, especially for the population. Manstein refers to the "rem lie," and states: "To use the term rem to characterize the effect on organisms is to either demonstrate one's ignorance with regard to the complex questions involved, or else to deceive one's audience."*[115]

*The reader will note rads and rems are often used mixed together. This should not cause confusion. For our purposes, 1 rad can be said to equal 1 rem. Moreover, the purely physical energy dosage rad is referred to as "absorbed dosage," while the biological-effect dosage rem is called the "equivalence dosage." It may therefore occur that both terms are used together, or that the word "dosage" alone may be used.

Artificial Atomic Fission

When uranium 235 is bombarded with neutrons, the atomic nucleus disintegrates into two parts. At the same time, two neutrons from the original nucleus are set free. If each of these neutrons in turn collides with a uranium 235 nucleus, new parts are again created. As well two neutrons are formed from each of the atoms, so that there are now four free neutrons. In this way, nuclear fission can reproduce itself independently, i.e., a chain reaction sets in. At the same time, high-energy gamma rays are emitted; heat energy is also liberated.

Thus, unimaginable quantities of atoms can be split in fractions of a second, as happens similarly in detonating an atom bomb. However, the fission process can also be controlled in nuclear power plants by using neutron-absorbing materials, like boron and cadmium, to direct the avalanche of neutrons. This permits a continual release of heat energy, which is then used in the usual manner to produce steam and electricity.

New Units of Measurement

We should not be confused by the new units of measure, which are greater than rads and rems by a factor of 100. Since 1985 the joule has been used as a unit of measuring energy quantity (calories are, of course, being converted to joules).

Thus: 100 rads become: 1 gray (Gy)
100 rems become: 1 sievert (Sv)

Initially, the clarity of the units is to a large degree lost, particularly for non-experts and the lay public. This book will in the main, be using the older units, rads and rems, which for our purposes do not need to be distinguished from one another. However, the new units are also given in parentheses.

The unit of measurement of activity, the curie, has been changed to the becquerel (BQ). One becquerel is present when one nuclear disintegration occurs per second. Thus, for example, 100 pCi (picocurie) is equal to 3.7 BQ (Becquerel). This, too, is irrelevant to us.

Exposure to Natural Radiation

We humans have always lived in a sea of natural radiation. No one can escape these rays. For three decades now, nuclear advocates have played down this threat. For instance, it cannot be determined if any harm has come to the population of Kerala, India as a result of the natural radiation dose, 1300 mrem/year, versus 130 mrem/year in the Swiss midlands, etc.[61] This leads to the impression that the entire range between 130 and 1300 mrem/year (0.13 - 1.3 mSv) is not harmful. However, natural exposure to radiation occurs both externally and internally, and it must be considered harmful.[68, 69]

External Radiation

External radiation has two sources: one is outer space—either from the sun or from remote bodies (cosmic radiation); the other is the earth.

The portion penetrating the atmosphere (primarily gamma rays, i.e., pure radiation) is weakened by the air, so that its strength depends upon altitude above sea level. A doubling of dose can be assumed for about every 1000 meters of altitude.

On the other hand, the earth's radiation comes from radioactive stones and minerals. The active elements potassium 40, uranium 238, thorium 232 and radium 226, which date back to primal times, as well as their successor substances, are responsible. Resulting radiation exposure can vary extremely, depending on the consistency of the soil. High values can be measured in certain areas of Brazil, Niue Island in the Pacific, and the Indian states of Tamilnadu and Kerala. Granite soils also cause higher radiation levels.

Internal Radiation

Internal radiation is the result of a very limited number of radioactive substances. These substances penetrate our bodies through respiration, as well as through the food chain. Not only the radionuclides from the ground have this capability, but also substances formed in the outer atmosphere. Such substances are radioactive hydrogen (tritium), and radioactive carbon (carbon 14), which are the result of transformation by cosmic radiation. The totality of internal radiation results not only in gamma radiation, but in alpha and beta radiation as well.

Overall Radiation

As early as 1966, the ICRP confirmed that natural radiation is harmful.[61, 68] The report noted that the majority of the world's population is exposed to a level 6 risk factor (1-10 deaths per million and per rad), while areas with higher natural radiation had risk factors of level 5 (10-100 deaths per million and per rad).

This means that in areas with high natural radiation, the risk is ten times as great as for the normal world population. In 1977, the ICRP stated that regional variation of natural radiation are thought to cause corresponding differences in damage.[69]

An example of normal natural radiation can be seen in the mean whole-body equivalence dose for a Swiss resident according to the Swiss Federal Commission for Radiation Monitoring in 1983:

External radiation:	
Ground radiation:	65 mrem/year (0.65 mSv)
Cosmic radiation:	32 mrem/year (0.32 mSv)
Internal radiation:	30 mrem/year (0.30 mSv)
Subtotal:	127 mrem/year (1.27 mSv)
Exposure from radon: (primarily in houses)	125 mrem/year (1.25 mSv)
TOTAL:	252 mrem/year (2.52 mSv)[101]

Artificial Radioactivity

Nuclear Power Plants—Only One Aspect of the Problem

The central danger of the nuclear fission industry is high-energy radiation and artificially-radioactive material. Both are increasingly harmful to our biosphere, due to the extension of nuclear technology, i.e., they will be the cause of ever far-reaching contamination.[8] These sources of danger accompany the entire process of nuclear industry, from the uranium mine through the production of fuel elements to the reprocessing facility and the temporary nuclear waste disposal—as of yet there is no permanent disposal. In addition, there is the transport of radioactive material. The entire fuel cycle or fuel chain has to be taken into account.

However, every new technology should now be tested for its possible extensive and long-term effects, no matter how economically valuable it may appear; this should occur prior to inception.

Nuclear advocates never tire of defining the nuclear power plant as the single aspect of radiation exposure to be discussed. However, the term "critical population group" has been introduced in reference to the population in the nuclear power plant's vicinity. In fact, the entire fuel chain (there is not yet in fact any cycle!) has long-range effects which can intersect and accumulate. For this reason, it is certainly possible for additional segments of the population to be strongly contaminated as well—particularly in cases of disturbances or accidents.

Atomic Bombs and Nuclear Power Plants

In the case of an atomic bomb, all the products of fission released immediately contaminate the environment, while a nuclear power plant emits "controlled" quantities of radioactivity into our habitat, either continually or in bursts. A nuclear power facility is never radiation-tight. In nuclear power plants, the same basic fission reactions occur as in a nuclear explosion. "Nuclear power fans" don't like to hear talk like that. Most of the fission products of a nuclear power plant are indeed retained in the spent fuel rods and filtration equipment, but now the insoluble problem of nuclear wastes arises. In addition, the reprocessing plants release much more radioactivity than the nuclear power plants, for many of the materials retained in the fuel rods are released only in these facilities. Included is 100% of the krypton 85, but also tritium, carbon 14 and iodine 129. The spent fuel rods, for instance, are transported to reprocessing facilities, where they are subjected to further processing for several purposes, such as the removal of plutonium.

Leaky Nuclear Power Plants and Environmental Contamination

A nuclear power plant is not like a flashlight, which no longer emits light when turned off. Rather, a nuclear power plant emits countless tiny glows in the form of artificial radioactive nuclides. These "flashlights" cannot be turned off. Their radiation wears off according to the physical laws of half-lives. Humankind can no longer intervene.

Myriads of invisible "flashlights" are emitted from the clean-looking smoke-stack, diluted by huge quantities of air (110,000 – 250,000 cubic meters per hour). These smoke-stacks are horrendous polluters. However, the emission also occurs through waste water and, as oftentimes in the past, in an uncontrolled manner through the machine plant and the ventilation.

Many construction materials are rendered radioactive in a nuclear power plant. This is caused by strong neutron bombardment. The corrosion products, which afterwards get into the water of the primary cooling cycle, include primarily the following isotopes:

Isotope:	*Half-life:*
Cobalt 60	5 years
Manganese 54	314 days
Iron 59	45 days
Zinc 65	245 days
Chromium 51	28 days

In spite of various filtration devices, a certain amount gets into the waste water as a result of leakage in pumps, slide-bars, valves etc., especially during repair work.

Finally both liquid and gaseous radioactive materials are formed by the neutron bombardment. This involves the impurities in the coolant water which appear in the primary coolant cycle due to disintegration of the water itself, seepage of air into the system, etc. For the most part, these impurities are the radioisotopes of nitrogen, oxygen, carbon (from carbonic acid), argon and hydrogen (tritium). The dangerous carbon 14 that is produced was only "discovered" in 1972. Prior to 1972, it had simply been ignored.[137] Nonetheless, we had always been told that everything was understood, and that everything was safe.

The multifarious mixture of small doses of highly dangerous radionuclides is randomly scattered throughout the environment, and hence, eludes all control. There is no such thing as protection against this radiation. These deadly artificial substances can enter the food chain without being seen, heard or felt. In this way, the infamous internal sources of artificial radiation are formed. In addition, the gaseous products affect our respiratory tract as well as our lung tissue.

The Potential Danger at our Front Doors

The main cause of the formation of "liquid" and gaseous emissions and wastes is the inevitable leaks (tiny holes and cracks) in the shells of the fuel rods. Through them, fission products leak into the primary coolant cycle and, partially, through its ventilation facility, into the vented air; in spite of all filters and retention devices. A 500 MW (megawatt) reactor may contain 30,000 fuel rods. The longer the fuel cells are in operation, the more the leakage rate increases. This is one reason why they have to be changed periodically. For this critical purpose, it is necessary to shut down the power plant.

After one year of operation (for a 1000-MW plant), the fuel rods contain a build-up of radioactive fission products equivalent to about 1000 Hiroshima-sized atomic bombs. This threat is at our doorstep, and its hazards exceed our imagination.

Concentration Mechanisms in Biological Systems

Consider this example of the extreme danger of radioactive contamination of food: A worker at the Hanford Plant in Washington State was diagnosed as having an inexplicable contamination of zinc 65. It was then ascertained that the man had eaten oysters from an oyster bed in the Pacific. Although the oyster bed was 400 kilometers away from the nuclear power plant, the zinc 65 had been carried in the waste water from the plant and concentrated there by a factor of 200,000.[138] Concentration factors of many thousands in living tissue are possible: up to 10,000 in fresh-water fish, up to 200,000 in plankton. The table below shows concentration factors that have been found for fission and corrosion products from nuclear power plants:[146]

Isotope	Sediment	Phyto-Plankton	Aquatic plants	Fish
Strontium 90	10-500	10-1000	10-10,000	1-200
Caesium 137	100-14,000	30-25,000	10-5000	400-10,000
Cobalt 60	400-29,000	—	200-24,000	400-4000
Iron 59	—	up to 200,000	up to 100,000	1000-10,000

In addition, such processes may also be reversible, so that it is therefore not safe to assume elimination of the radionuclides from the waters. The transportation of suspended matter is a major factor.[147]

Thus, unexpected mechanisms may result in nature, the food chain, organs and organic systems. This may lead to specific effects on organs and risks which have yet to be explored thoroughly.[41]

Concentration in such sensitive organs as bone marrow, hormone -producing glands and also germ cells, embryos, etc., by way of inhalation or ingestion, can cause doses from internal radioactive sources of factors of 10-100 times larger than what they would have been if these isotopes were having their effect from outside.

Vegetables grown in the vicinities of nuclear power plants can also be transported elsewhere, so that a consumer far away from the facility may be exposed to more hazardous radiation than a neighbor of the plant who eats something different.

Obviously, there is also the risk of contamination of farm crops watered by surface water or fertilized with treated sludge; contamination of milk with iodine, strontium, cesium, etc., is a given.

These mechanisms show that too much is unknown, and that the emission of radioactive materials from nuclear facilities is in no sense simply a matter of physical dilution. It should be remembered that our supply of drinking water will in the future be ever more dependent on surface waters, into which the waste of nuclear power plants flow.

The problem can never be brought under control. The radionuclides produced by our technology can find their way through the complex food chain from one creature to the next, and from one generation to the next. Sometimes they will concentrate, sometimes disperse. They keep causing more and more damage. No one can predict where and when these radioactive emissions will turn up in the air or water—or on our plates. Depending on their properties, these emissions will still be around for years, decades, centuries, and, as in the case of iodine 129, even for millions of years.

An Overview of Radiation-Protection

Radiation-Protection Authorities

Anyone who would like to discuss nuclear power or form an informed opinion should acquire a general understanding of the possible biological and medical effects of radiation and radioactivity. This is particularly true of the tragic phenomenon of forest death.

Crucial to understanding nuclear power is the examination of behind the scenes radiation-protection rule-making. The information necessary for the calculation of radiation-related risk is for the most part still uncertain. Many of the statements which follow are based on publications of the following offices: the International Commission for Radiological Protection (ICRP); the U.N. Scientific Committee on the Effects of Atomic Radiation (UNSCEAR); and the U.S. National Academy of Sciences' Biological Effects of Ionizing Radiation (BEIR).

Most of these reports confirm our ignorance and do not allay our well-founded fears. The offices in question, moreover, are under the influence of both the nuclear industry and pro-nuclear government authorities. The radiation-protection laws of all nations are based on the recommendations of these three organizations, particularly those of the ICRP.

Fundamentals of Radiation-Induced Damage

We distinguish between the following two effects of biological radiation-induced damage:

Somatic damage: Damage caused to the irradiated person him/herself.

Genetic damage: Genetic damage which becomes evident or destructive only in later generations.

Biological damage begins at the zero dose level, as UNSCEAR has stated:

> Studies of relationships between dose and effect at the cell level or below offer no indication as to a level of tolerance, which nec-

essarily leads to the conclusion that biological damage takes effect after radiation, no matter how small the dose is.[195]

Furthermore, they state that in order to determine whether an immediate or a delayed effect is to occur a number of factors must be considered. The key factors include, the individual radiation-sensitivity of the person. Radiation-protection legislation pays too little attention to this factor, so that sensitivity is sacrificed from the outset. The vicinity of nuclear power plants includes sick people, pregnant women and babies, who are particularly vulnerable to radiation. Nonetheless, these people have not received any special protection in radiation-protection legislation. Even the ICRP states this clearly as recently as 1984.

> There are a number of additional factors which need to be investigated. . . . Pregnant women and the chronically ill should receive special attention. More should be known about the metabolism of radionuclides (biological behavior) in the embryo and the fetus, and about their susceptibility to radiation.[93]

Among the most important effects of radiation is that on cell growth—the cell nucleus, the cytoplasm and the cell wall. Every organism is composed of trillions of different cells. The growth of biological matter occurs through the reproduction or splitting of tissue cells. However, there are many tissues in mature bodies which are constantly renewing themselves more or less rapidly through the death of old cells and the formation of new ones (e.g., blood and sperm cells, the skin and the mucous membranes).

Ionizing radiation affects the structure and chemistry of the cells. Ions are formed, and molecules may be split apart, forming radicals. Radicals are pieces of molecules that are chemically very aggressive, and that can form new compounds which are foreign and in some cases, toxic.

It has long been known that radiation does not damage cells equally, but particularly targets those cells which are rapidly multiplying. This is why medical science uses ionizing radiation in order to destroy cancer cells. However, a certain risk is associated with this treatment, since healthy tissue is also irradiated. Nonetheless, it would be a mistake to avoid or refuse vital radiological examination or treatment out of a misguided fear of radiation; moreover, physicians have become much more cautious in this regard.

Genetic Damage

Dominant and Recessive Damage

Genetic damage occurs as the result of changes (mutations) of the chromosomes contained in the cell nucleus. The basic structure of chromosomes is the threat-shaped deoxyribonucleic acid (DNA). The DNA undergoes a complex but very precise transformation of form and function during the various phases of multiplication and differentiation of the cells. The species-specific development potential of an individual is determined by the totality of his or her genes. The DNA is the carrier of this genetic information. The human egg and the human sperm each transmit 23 chromosomes. Thus, the fertilized egg contains 46 chromosomes. As the result of the specific molecular construction, each of the 46 chromosomes is reproduced prior to the cell division. During cell division, a complete set of 46 chromosomes is then transported into each of the resulting daughter cells. Identical reproduction of DNA and equal distribution of chromosomes are the indispensable preconditions for the normal development of an organism.

Any structural modification of the chromosomes of a germinating cell, including that caused by radiation, leads to a change in the genetic information passed on to descendants, i.e., to a genetic mutation. If, on the other hand, a normal body cell is affected, the descendants will not be damaged—only the individual exposed to the radiation. This so-called somatic mutation may, for instance, result in cancer.

On the other hand, genetic damage in the cell nucleus may be repaired.[52] This particularly protects the genetic information of life forms from natural radiation. However, there are certain substances which may also inhibit the repair capacity (e.g., caffeine).[52] The marked differences in radiation-sensitivity in the population may be linked to differences in the degree to which radiation damage can be repaired.

Gregor Mendel described two fundamentally different types of genetic heredity as early as 1875—dominant and recessive.

- A dominant mutation causes all carriers of a mutated gene to display a modified, possibly pathological, trait.
- A recessive mutation is in some ways more deceptive. It may be passed on unmanifested until the day when an egg and a sperm

carrying the same recessive mutation join during fertilization. This may take many generations. The carrier of a recessive mutation does not necessarily notice any effects due to the concealed genetic damage. An apparently normal picture of a healthy population may be misleading. In the future, when a critical distribution level with a possible high combination of damaged partners is achieved, the harm inflicted on the population will appear as a genetic catastrophe.[208] Recessive mutations occurring today will be affecting future generations centuries or millennia from now.[57, 65, 198, 204] There are intermediate stages between the recessive and the dominant mutations, but all mutations are as a rule harmful, and cause a reduction in vitality and fertility.

Obvious deformities, such as congenital disability, or such rare diseases as hemophilia and dwarfism are not the primary results of genetic damage. These congenital conditions, which are in general fairly well understood, are only the tip of the iceberg.[13] Still more serious, virtually all diseases may have a genetic component, and may therefore be impossible to cure. The list of such diseases includes allergies, epilepsy, arthritis, kidney disorders, diseases of the liver, mental retardation, eye diseases, muscular dystrophy, bone disease, degenerative diseases of the brain, arteriosclerosis, heart diseases, and the loss of vitality.

The apparent ease with which most radioactive substances can be detected (but not, by a long shot, all—e.g. tritium), and the apparent exactitude of the admissible doses, scenario calculations, etc., can give a completely false sense of security.

There have always been natural mutations. These are called spontaneous mutations, which include those radiation-induced mutation rates caused by natural radiation. We should therefore get used to the idea that spontaneous mutation, too, can cause genetic and other health-related damage, such as cancer. However, even today, only vague estimates are possible regarding the genetic and health-related negative effect of natural radiation. Scientists have widely different opinions in this regard. BEIR III 1980 has estimated that between one and six percent of the spontaneous mutation rate can be attributed to "background" (or natural) radiation.[23] Archer believes that 40% to 50% of the spontaneous cancer rate is the result of natural radiation.[2] There are also studies which indicate that deformities of the human population are more common in areas with strong natural radiation than in those with average background radiation.[55]

It is also possible for artificial radiation, certain chemicals and medications to cause mutations. Unfortunately, the long-term effects of many chemicals in our lives have yet to be thoroughly studied. Humanity is handling its genetic heritage rather carelessly. In addition, it is a great deception to claim that radioactivity is better understood than all other poisons. Nuclear energy advocates would have us believe that we have a handle on radioactivity, which is neither true nor even much comfort, unless we are eager to check ourselves and our food daily with Geiger counters and radiochemical tests. On the contrary: artificial radioactivity is emitted at random from nuclear facilities, and is thus removed from any control.

Today, no authority responsible for radiation-protection recognizes any level of tolerance in regard to genetic damage, and the mutation rate of genes increases linearly with the radiation dose, starting at zero (see Figure).[208]

The number of induced point-mutations (here, lethal factors in *drosophilia*) is proportional to the radiation dose (from Timofeyev-Resovsky et al., 1972; cited by numerous authors).

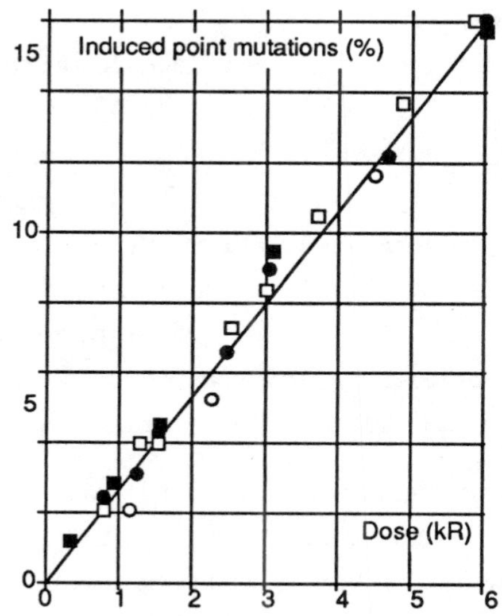

Maximum Permissible Doses

Though H. J. Muller had discovered that ionizing radiation causes mutations as early as 1927, and hence was genetically damaging, dose limits for the population of the world were not established until the beginning of the atomic age. Until then, only limited groups of people had been exposed to artificial radiation, mostly doctors and patients using X-rays. As a result of nuclear weapons tests, since 1945, and the introduction of nuclear power, all of humankind has been contaminated with artificial radioactivity.

This is why in 1956, the United States National Academy of Sciences came up with the idea of using the natural radiation level as a measure of the maximum permissible does for the world population. A mean level of 5 rads (50 mGy) in 30 years, or 170 mrad (1.7 mGy) per year was assumed. In effect, a doubling of the natural level of radiation was believed to be acceptable.[21,52] The genetic damage from weak radiation was assumed to be slight.

In 1958, therefore, the International Commission on Radiological Protection (ICRP), whose recommendations serve as a model for the radiation-protection regulations of all nations, established a maximum permissible dose for the world population of five rem in 30 years, or 170 mrem per year (1 rem - 1000 mrem). This limit did and does not include medical and natural radiation exposure.[24]

Genetic considerations had been the exclusive concern. The belief was that there would be no risk of cancer at such low radiation doses (the level of tolerance). American statistics were used to show that on the average, children were conceived by parents up until 30 years of age. For this reason the 30-year period, rather than lifetime exposure, was used for the calculation. Manstein remarked: "This arrogant judgment contains an obvious threat: woe be to those who defy the norm and have children later in life; artificial radioactivity will have no regard for them."[114]

It should be noted that the ICRP made the following comment:

> The Commission believes that this level would provide reasonable latitude for the expansion of atomic energy programs in the foreseeable future. It should be emphasized that the limit may not in fact represent a proper balance between possible harm and

> probable benefit, because of the uncertainty in assessing the risks and benefits that would justify the exposure. (ICRP 9).[69]

Thus, dose limits were not established for the purpose of protecting the public, but rather in order to provide nuclear power with reasonable latitude for expansion. At the same time, a maximum dose for occupationally-exposed persons of five rem/year (50 mSv/year) was established (here, too, the genetically based 30-year period applied).

It is extremely important to note that these doses have not been reduced, even though low-dose radiation has since been shown today to be 100 to 1000 times more harmful.

Above all, the risks of natural radiation can no longer serve as an alibi. Genetic information must be valued as our most precious possession, and any artificial violation of it by nuclear power—be it by way of cost-benefit analyses, risk analyses or whatever—should be branded as a heinous crime.

The subcommittee of the National Academy of Sciences stated in its BEIR III Report of 1980:

> . . . The subcommittee is convinced that any increase in the mutation rate will be harmful to future generations.[22]

In 1972, the BEIR report had stated that, with the development of nuclear power it is inevitable that the biosphere will be subjected to increased radioactive exposure.[8] In plain language: We will just have to take into account increased mutations and genetic damage.

No More Advanced than Thirty Years Ago 31

In spite of revolutionary discoveries in the area of genetic damage—including the discovery of the genetic code—medical radiobiology has no better quantitative data than it had in 1952. The 1980 BEIR report merely repeated what had already been said in the 1972 report.[9, 20] Although the committee had begun to use a new method of risk estimation for the first generation, estimates of genetic effect were hardly different from those in the 1972 report.

Thus, when required to provide concrete data, medical science found itself in exactly the same helpless position it was in over thirty years ago. It is no wonder the well-known radiobiologist Alexander Hollaender describes the development of his branch of science as a "field of lost battles."[46]

The gaps in our knowledge are serious. The 1972 BEIR report exposes them ruthlessly, citing "serious gaps in our knowledge" regarding radiation-induced mutation rates in humans.[10] It adds:

> It is impossible to prove that the doubling dose is not as low as the background radiation level [natural radiation exposure], namely about three rem.[14]

Of course, it is assumed that this doubling dose (i.e., the dose that would double the natural or spontaneous mutation rate) is somewhere between 50 and 250 rem. This assumption is based on the uncertain transfer of results from experiments on mice to human beings.[25]

Even if we were to know this doubling dose, or the increased mutation rate per unit of dose, we would not be much further along. The 1972 BEIR report notes: ". . . our inability to quantify the relation between an increased mutation rate and deleterious effects on human well-being."[10]

In other words, we know virtually nothing. What we do know is largely the result of experiments upon animals, particularly mice and fruit-flies (drosophila) bombarded with high doses of radiation. However, it is known that genetic results cannot be transferred from one animal species to another with any degree of accuracy—let alone from an animal to humans.[10] The same BEIR report also points out that "by relying so heavily on experimental data in the mouse, we may have overlooked important effects that are not readily detected in mice."[13]

Elsewhere: "We can't ask a fruit fly whether or not it has a headache."[12] Obviously, the same is true of mice.

What has been found in drosophilae is that minor mutations occur ten times as frequently as serious, lethal mutations.[11] It had previously been thought that minor mutations exceeded the serious ones by a factor of only two or three.[11] The former mutations are the more deceptive, since they have a lesser effect on fertility and vitality. Therefore they can be transmitted all the more easily to future generations.[12]

The 1972 BEIR report stresses the extreme significance of these minor mutations that manifest themselves in animal experiments:

> Perhaps the human counterparts of these mutations, in addition to causing a slight reduction in life expectancy, are responsible for greater susceptibility to disease, impaired physical or mental vigor, or slight malformations of some organ.[12]

Nuclear advocates prefer to point out that the dangers of radiation are exaggerated by citing mice which have been subjected to high radiation doses without showing any visible damage for 40 generations.[10] But, in this process, minor mutations may have been overlooked—as noted above, it is impossible to ask a mouse whether it has a headache, or find out if its mental capacity has been reduced—and the experiments may not have been broad enough. Just as obviously, any reduction of instinctual behavior would be apparent only when a lab mouse population had been released into the wild, and subjected to the rigorous conditions of natural selection.

The 1972 BEIR Report has the following to say regarding such experiments with mice:

> Yet there is the possibility that one does not see in mice effects that would cause appreciable distress in humans.[14]

Professor Hedi Fritz-Niggli, Chairwoman of the Radiobiological Institute of the University of Zurich, Switzerland, which supports nuclear power, says in regard to mutation induction:

> Risk estimates according to USCEAR 1977, ICRP 26, 27 and BEIR III 1980 are based on findings from animal experiments (particularly mice). Since the genetic mass of higher life forms should not react to radiation species-specifically, the transfer of experiments from small rodents to humans is certainly justified.[53]

Commentary in June 1981, by the Swiss Federal Expert Commission on Dose Effect, in response to the 1980 BEIR Report (chaired by Fritz-Niggli), summarizes the "Effect of Small Doses of Ionizing Radiation on the Population" in terms of genetic damage as follows:

The values of 5–75 manifest, genetically-determined disturbances (in addition to the 107,000 'spontaneously' occurring anomalies) per million live births after exposure to one rem, which were obtained by BEIR III (1980), appear to be valid as genetic risk estimates, as are the revealing ascertainment of 60 - 1100 genetically-related disturbances after continuous exposure of each generation to 1 rem. The 'natural genetically-determined occurrence of anomalies was at that time raised by 0.05% to 1.03%. (1rem = 10mSv)[48]

The 1972 BEIR Report does in fact state that the well-researched genetic diseases may form only the tip of the iceberg:

> What about the rest of human illness? It, too has some degree of genetic determination . . . major concern of the Subcommittee is the possible existence of a class of radiation-induced genetic damage that has been left out of the estimates. By relying so heavily on experimental data in the mouse, we may have overlooked important effects that are not readily detected in mice, or the mouse may not even be a proper laboratory model for the study of man.[13]

Back in 1966, the ICRP had noted with great honesty:

> Because the total genetic detriment will become manifest only over very many generations, it is appropriate that long-term effects should be a major preoccupation of the collective conscience; from this viewpoint the 'total eternity damage' may be considered.[65]

That same year, UNSCEAR had been just as honest:

> For most genetic changes, even conjectures are not permissible regarding the actual manifestation of the damage throughout generations in terms of individual or collective hardship.[198]

More recently, a UN Report (UNSCEAR 1982) states:

> The problem of the contribution of recessive mutations, both spontaneous and induced, to the human genetic burden has been and continues to be one for which it is difficult to provide reliable quantitative answers.[204]

The sum of these quotes cited from the three most authoritative bodies on radiation-protection and radiobiology demonstrate the inadequate state of modern radiobiology with regard to long-term genetic damage by nuclear fission. In other words, the dose levels being established today offer no secure protection to future generations. Even the radionuclides emitted from nuclear power plants in uninhabited areas do ultimately enter our habitat. The soil, the seas, the atmosphere and the entire biosphere are thus unremittingly being strongly contaminated. Given this situation, how can nuclear power plants and reprocessing centers be defended, let alone labelled as environmentally benign?

Yet, for Fritz-Niggli, all this appears not to represent a problem. She has calculated that Chernobyl will cause only between two and 22 mutation-generated anomalies in Switzerland over the course of the next century.[217] By contrast, according to the noted scientist and author, Dr. P. Weiss of the Institute of Environmental Science and Natural Protection of the Austrian Academy of Sciences, the Chernobyl catastrophe threatens future generations with consequences of continental proportions due to genetic damage.[216]

Damage to Health (Somatic Effects)

In contrast to genetic damage which affects our descendants, somatic damage occurs in the radiated individual him or herself. For this purpose, it is useful to distinguish between the effects of high and medium doses (acute damage apparent shortly after irradiation, such as reddening of the skin), and those of low doses (delayed effects). However, there are also delayed effects after one recovers from acute radiation disease. Moreover, since the discovery of the Petkau Effect in 1972, the possible consequences of cell membrane damage, serious with even the most minimal doses, are also a considerable threat.

Effects of High and Medium Radiation Doses (Acute Effects)

Short-term death resulting from accidents and catastrophes at nuclear power plants are indeed gruesome. In the case of a failure of the emergency cooling systems, these deaths could occur by the thousand; in the case of a nuclear war, they would be in the millions.

Acute radiation damage as the result of high or medium-dose whole-body irradiation was first experienced by human beings after the atomic bombings of Hiroshima and Nagasaki in 1945. In addition to burns and injuries, the victims showed various symptoms of acute radiation syndrome, which varied depending on their location and individual susceptibility to radiation. These individual syndromes, which manifested within hours or days after the damage, could also be attributed to other diseases. The victims suffered from headaches, dizziness, vomiting, fever, diarrhea and apathy, and often died within a few days. Moreover, a few weeks later, seemingly uninjured persons also became ill: they vomited blood, developed bloody spots all over their bodies, and urinated blood; later, they would start to lose hair, find blood in their feces, and feverous infections due to a lack of white blood cells would set in. The survivors for the most part faced a lifetime of wasting illness. Nuclear accidents that have since occurred offer the opportunity to confirm this picture of acute radiation syndrome.

A single whole-body irradiation causes approximately the following reaction: between zero and 25 rems, no effect can be detected; between 25 and 60 rems, 10% of victims react with nausea and vomiting; at 180 rems the first fatalities begin to appear, and 25% of irradiated victims will suffer from radiation sicknesses. At 300 rems, fatalities will amount to 20%; in the 420-700 rem range, they reach 90%. At doses above 1000 rems, survival is improbable, and death occurs within hours. It should be noted, however, that the data is by no means uniform.[131]

Effects of Low-Level Radiation Doses (or Delayed Effects)

The ominous effects of radioactivity were first noticed in the last century (1878) during iron ore mining in the poorly-ventilated Schneeberg mines in Saxony, eastern Germany. Here the rocks held high concentrations of uranium. Cancer of the lung took the lives of 75% of the workers. Today we know that not only the dust containing uranium and thorium was responsible for these deaths, but also radon, a radioactive inert gas which was concentrated in the air of the mine. Around the same time, the pioneers of the X-ray would carelessly place their hands in the radiation beam to determine its intensity. Soon, the first effects of radiation would appear on their hands. In 1902, the first cancer of the skin following X-radiation was described. Other important historical dates include:

1911: Publication of the first paper on X-ray-induced cancer (latency period: 9 years), by O. Hesse.[33]

1911: First reports of leukemia among doctors and nurses using X-rays.[33]

1928: Thorotrast is used for the first time as a contrast agent in X-ray diagnosis. Only much later will it be known that these patients had come down with leukemia and cancer. According to a study conducted by the German Institute for Cancer Research in Heidelberg, current as of 1982, cases of cancer of the liver exceeded those of leukemia by a factor of 10. The shortest latency period for leukemia was 5 years; for cancer of the liver it was 16 years.[40]

1929: Women workers who painted luminous numbers onto the dial-faces of watches with fluorescent paint containing radium come down with bone cancer.[33]

1930: It is experimentally proven that leukemia is caused by radiation.

1956: First studies are undertaken by Alice Stewart on cancer in children after diagnostic X-ray examination in utero.[33]

1957: Only at this point is it realized that cancer would occur in survivors of the atomic bombings in Japan even many years later. Prior to this, leukemia had been virtually the only disease to arise. There was a feeling of certainty that this was the most prominent radiation sickness, and that the risk of cancer was negligible.

1960: According to former ICRP Chairman Karl Z. Morgan, virtually all scientists believe (until 1960) that there is such a thing as a level of tolerance, i.e., a maximum dose which would still be harmless. As long as this dose is not exceeded, it is believed, no damage to health should be expected. (Since 1960, however, an overwhelming flood of data has shown that there is no such thing as a level of tolerance for radiation-induced cancer.[120]

1966: The ICRP stubbornly continues to hold to its position that all is well:

> . . . doses of the order of a few hundred rads are capable of causing cancer in certain tissues. No evidence is available (or likely to become available) for the effect of smaller doses on organs other than the marrow of the adult. . . (100 rad = 1 Gy).[66]

Diseases caused by radiation may appear years or decades after acute radiation sickness has been overcome. The same is true for lower exposures to radiation—which strike over a longer period. However, low-level radiation does not cause any specific radiation sickness (radiation syndrome); rather it fosters a large number of ordinary diseases, which will appear to be statistically increasing. Furthermore, not all victims of irradiation will get sick—only some. It is like Russian roulette, where no one knows whether he or she will be the one to be hit.

A large number of diseases can be caused by low-level radiation:

- leukemia;
- cancer of all types;
- reduced fertility;
- chromosomal changes of the blood;
- physical and mental defects of the unborn (developmental damage);
- genetic damage.

The following possibilities are also being explored:

- disruption of the hormonal and enzyme balances;
- increased susceptibility to infectious diseases, and to heart and circulatory diseases;
- premature aging;

It should be emphasized that radiation increases a large number of health hazards.[214]

Of the above-listed disease risks, leukemia and cancer have been the most thoroughly investigated to date. Even in the case of these ills we

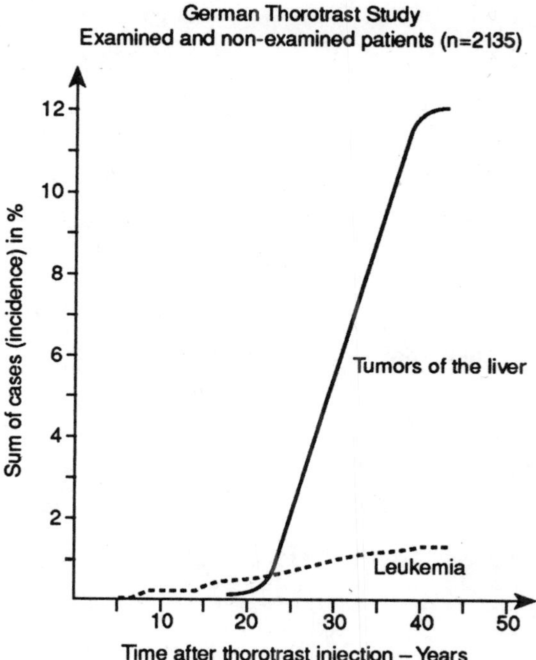

are a long way from being able to make serious estimates. It is believed, that cancer may be the result of a somatic mutation in the cell nucleus of a body cell. Thus, any cell may mutate to form a cancer cell. Today we still do not know the exact mechanism. Since each high-energy interaction will have a certain likelihood of causing such a transformation, no threshold value (harmless dose) may be assumed for theoretical reasons. This means that, analogously to genetic damage, a portion of the spontaneous cancer and leukemia cases have to be attributed to natural radiation, and that any additional radiation exposure due to nuclear power and medical radiation will increase the rate of cancer and leukemia.[208]

In 1970, low-level radiation doses were shown to cause cancer in the very radiation-sensitive bodies of unborn babies. For instance, if babies in their mothers' wombs are X-rayed, they have an increased risk of cancer through their 10th year, according to Stewart and Kneale, 1970.[108]

Number of X-Ray Exposures (0.2 - 0.46 rad per null exposure) (2 mGy - 4.6 mGy)	Increased Cancer Rate in %
0	0
1	20
2	28
3	70
4	100

In 1982, a long-term study was published which showed that women irradiated in their mothers' wombs were subject to a 5.5 times greater risk of cancer of the breast at the ripe age of 30.[62]

Tokunaga et al. have reported an increased rate of breast cancer in Japanese women who were irradiated by the atomic flash at age 10 or less.[188]

The Radiation-Protection that No Longer Is

Dependent Authorities

Since the 1970s, it has become increasingly evident that ICRP and UNSCEAR authorities feel obliged to hold open the doors for nuclear power. Their position is hardly surprising, despite the massive increase in radiation risks.

The members of the U.N. Scientific Committee on the Effects of Atomic Radiation are appointed by the governments of the nations concerned. This generally means that only pro-nuclear scientists can secure these positions.

At the International Commission on Radiological Protection, too, the method of selection ensures which scientists will participate. Former ICRP Chairman Karl Z. Morgan himself, has criticized this.[215] Publication No. 26, 1977, shows that the ICRP is financed by a list of bodies which all support nuclear power, including:

- The World Health Organization (WHO);
- The International Atomic Energy Agency (IAEA);
- The UN;
- The International Radiological Society;

- The International Radiological Protection Association (IRPA);
- The Nuclear Energy Agency (NEA);
- The European Community;
- various unidentified national sources in Canada, Japan and Britain.

The commission confirms, albeit unintentionally, its financial connections:

> The Commission is grateful for the time which its members were permitted by their institutions to devote to this work, and for the financial support without which it would not have been possible to conduct this work.[87]

The American National Academy of Sciences also issues key reports. For example, they are responsible for the so-called Biological Effects of Ionizing Radiation Reports (BEIR). One famous report was that of 1972, which provided a good, objective overview of the then-current state of knowledge, written in an easily comprehensible form; in fact, as we shall see later on, it was a pioneering work (for instance, in regard to risk calculation). In 1979, the BEIR III report was published, and it, too, was surprisingly frank. Too frank, as would soon be shown, for it was quickly withdrawn. In its place, there appeared a revised 1980 BEIR III report with data that played down the danger. This classic example of manipulation of important scientific data will be discussed in greater detail.

Gofman and Tamplin: The First Independent Risk Calculations

The most important events of 1969-1970, as far as the issue of radiation is concerned, were the groundbreaking publications and statements of the researchers John W. Gofman and Arthur R. Tamplin of the University of California at Berkeley.[58, 59, 60] The U.S. Atomic Energy Commission had permitted an additional dose of 170 mrem/year (1.7 mSv) for peaceful utilization of atomic energy. According to Gofman and Tamplin, full utilization of this dose would mean between 16,000 and 30,000 additional deaths from leukemia and cancer per year, over the course of 30 years in the U.S.A. alone. They demanded a reduction of the dose to one tenth of that figure, i.e., to 17 mrem/year (0.17 mSv).

What is ironic about the study is that it was carried out under contract, and with the support of, the AEC. The work, which was begun in 1963, was designed to check the dangers of radioactive emissions to the biosphere and to human beings. Fortunately, the two scientists had the courage not to shrink from making their serious findings public, and did not simply submit a report tailored to the desires of the Commission. The pro-nuclear AEC then showed its true face, and proceeded according to its usual method of dealing with scientists, experts, Nobel-prize winners and any other outsiders who opposed nuclear power: Gofman and Tamplin were slandered and ridiculed. They were threatened with dismissal, their material was censored and modified. Their staffs were reduced, their salaries were cut, and they were refused raises. In spite of it all, the AEC was not able to refute the researchers. Even their superior was unable to pinpoint any errors. Interestingly, the AEC avoided a public debate, as the two scientists suggested.[182]

Radiobiology is today still in its infancy. Seen from the point of view of radiation-protection, its development to date has been one long drama. Since their discovery, radiation and radioactivity have repeatedly been found to be more and more dangerous. For some reason, this fact continues to be denied. We should not let ourselves be fooled by the authorities of countries, including their agencies for radiation-protection, nuclear safety, etc, and also the nuclear industry, who constantly trumpet the harmlessness and environmental soundness of nuclear power. Official scientific literature (ICRP, UNSCEAR, BEIR, etc.) has found it more and more difficult to cover up the previously understated dangers of radiation—since they were, depending on the particular case, higher by factors of 10, 100 or even 1000.

This attitude of the authorities—from which the nuclear industry continues to profit today—was originally a product of the Cold War. The Western and Eastern nations concerned about their security, were not willing to let themselves be constrained in the expansion of their nuclear arsenals and the conduct of their nuclear tests. Their citizens were insufficiently and falsely informed of the dangers of radiation and radioactive emissions. In his book *Secret Fallout*, Professor Ernest J. Sternglass, on the basis of documents which were only made public years later, confirms this censorship.[177] These documents inform the public that atomic bomb explosions basically release the same dangerous radionuclides as nuclear fission from a nuclear power plant. Delayed publicity of this danger was to maintain the widespread idea of a harmless level of tolerance.

This indicates the extent to which the professional dependence of scientists influences their statements, and how unwise political decisions can be made if politicians have a one-sided faith in dependent experts.

Thus, for instance, the Swiss Association for Atomic Energy (SVA) published a full-page article in the August 26, 1970 edition of the *Neue Zurcher Zeitung*, Switzerland's leading newspaper, stating that the recommendations established by the ICRP would "have no effects on health or genetic damage." The same newspaper in 1972 played down the danger of radiation from maximum admissible doses with the statement: "Of course, not one single case of illness or death other than from natural causes may be tolerated."[126]

By contrast, in 1966, the ICRP had already stated the following:

> The foregoing makes it apparent that this report can provide no simple solution to the practical dilemma in setting criteria for radiation-protection. Although quantitative recommendations are required for the control of the nuclear energy industry and for the protection of the population in emergencies, the evidence on which such recommendations can be based is imprecise.[67]

The evidence is not just imprecise; decisive elements did not and do not even exist. In any event, the scientists were at that time honest enough to admit a dilemma. Apparently they were forced to blaze a trail for the nuclear industry in the teeth of their own misgivings. As time went by, expressions of concern were no longer voiced. Governments had placed a tight rein on the commissions—particularly the ICRP and the UNSCEAR.

Thus, the ICRP publication of 1969 cynically directed that calculations should be made as to how many victims in the population would be "acceptable." The idea of risk acceptance in radiation-protection, which contradicts all morality and ethics, was born. There, it is stated:

> Where tumors and genetic effects are concerned, it is generally postulated that no threshold exists. . . Protection recommenda-

tions now have to be designed to reduce the total probability in a population to an acceptable limit.[73]

The population was not informed of this until after the fallout at Chernobyl. The oft-repeated slogan still is: "No one has ever died because of radiation from a nuclear power plant."

Early European Warnings

Gofman and Tamplin gave this author a clarion call. I had already become a convinced opponent of nuclear power after studying the ICRP and UNSCEAR publications, which showed that radiation-protection legislation was based on insufficient foundations—affirmations to the contrary by authorities and pro-nuclear scientists notwithstanding.

In Germany, Austria and Switzerland, too, there were already prominent opponents of nuclear energy. The list includes Bechert, Bruker, Heitler, Herbst, Manstein, Niklaus, Par, Scheer, Schwab, Schweigert, Thierring, Thürkauf, Weish and Zimmermann. Professor G. Schwab, founder of the World Association for Protection of Life, deserves special mention for arousing the public as early as 30 years ago with his book *Der Tanz mit dem Teufel (The Dance with the Devil)*.[157] Without the Club of Rome's computer, this great thinker had already discovered the threat to the environment, and had analyzed and confirmed it. The special interest groups fell upon him and tried to force this "troublemaker" to withdraw his book.

Unfortunately, at that time there were no books regarding the dangers of nuclear power which were based upon the radiobiological literature of the ICRP and the UNSCEAR. After my having failed, through newspaper articles and letters to the editor, to alert the public to the dangers and manipulations regarding nuclear power (the media for the most part rejected these pieces), my book *Die sanften Morder—Atomkraftwerke demaskiert (The Gentle Killers: Nuclear Power Stations Unmasked)* appeared in 1972. It was translated into a number of languages and became a best-selling paperback in 1974.[61]

Nuclear advocates reacted quickly. Soon, critical newspaper articles began to appear, as well as a privately published pamphlet. The author was the physicist H. Brunner, secretary of the Swiss Professional Association for Radiation Protection.[36] He writes:

> Gofman and Tamplin are using milk-maid's arithmetic. Like Graeub, they claim that it is both officially permitted and possible to expose the population to 170 mrem per year due to the operation of nuclear power plants. From this, they concluded that this would result in 32,000 additional deaths per year in the U.S. due to cancer and leukemia. They therefore call for a reduction of the dose to one tenth.—Where does Gofman get the moral justification to be satisfied even with one tenth of that—3200 deaths—if he really does believe in his calculations?[35]

At the end of 1972, the BEIR I report was issued by the U.S. National Academy of Sciences, basically confirming Gofman and Tamplin's study. The report in effect demands that risk calculations for cancer should in the future serve, together with genetic risks, as the basis for radiation-protection legislation for the general public. Also, in general agreement with Gofman and Tamplin, an additional 15,000 deaths per year from cancer and leukemia were estimated.[19] Thus, the two scientists were rehabilitated, and their critics embarrassed. Soon thereafter, Gofman and Tamplin demanded a reduction from 170 mrem (1.7 mSv) to zero, while the Academy of Sciences recommended a reduction to only a few mrem.[7]

The "Bible" of Radiation-Protection Legislation

International radiation-protection bodies like the ICRP, UNSCEAR, and BEIR based their information regarding somatic damage (i.e., risks of cancer and leukemia) largely on investigations of survivors of the atomic bombings of Hiroshima and Nagasaki, still alive in 1950. Thus, some 80,000 Japanese radiation victims were observed and the causes of their deaths evaluated. Each person was subjected to a certain radiation dose, which had to be estimated on the basis of his or her location at the time of the nuclear explosion. This estimate was called the TD65 Study, and was based on a short-term exposure to radiation—the explosion.

Dependent on the dose which the survivor had been exposed to, delayed effects such as leukemia and cancer, have appeared over the years. Since it is impossible to distinguish between radiation-induced and spontaneously occurring cancers, it is necessary to fall back on statistical studies. In such studies, the number of cancer cases in a population which has been subjected to a known radiation dose is compared with the number of cancer cases in a non-irradiated population. This so-called control group should be sociometrically as identical as possible with the irradiated group. At high doses, the increased number of cancer cases in the irradiated group is clearly ascertainable. In the case of low doses, however, this increase may be so small that it can no longer be detected with any statistical certainty. This is due to fluctuations produced by natural causes. In order to get an estimate of the risk, it is necessary to calculate mathematically the effect of small, low doses, based on the observed effects of higher doses, in accordance with the best existing professional opinion. In other words, low-level dose effects are derived from the results observed at high doses. This can be represented graphically by means of so-called dose-effect curves. The higher the dose, the more cancer cases are to be expected.

Dose Effect Curves

These curves are based on complex arithmetical calculations. The graph shows several typical forms to illustrate in a simplified manner the various extrapolations of the effects of high doses to smaller ones. They show both how great our radiobiological ignorance is and how well-founded our fears are.

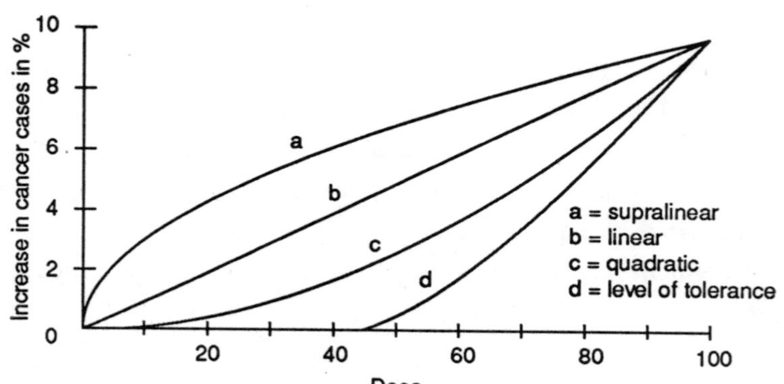

The vertical axis shows that the percent increase in cancer cases—from zero to ten percent over the normal incidence—caused by the radiation doses indicated on the horizontal axis (from zero to 100 arbitrary units of radiation). In the above arbitrary examples (curves a through d), it is assumed that practical investigation has shown that a high dose of 100 units causes ten percent more cancer cases than in a control population (see the point of intersection of the four curves at the top right of the graph, which is also the intersection of the lines for "dose of 100" to the left, and for "ten percent" below). To discover what will happen at lower dose rates, which for the above described statistical reasons may not be ascertained experimentally, the line may be extended toward zero by means of various curves. Four examples are shown, with each curve yielding different results:

b. The straight-line curve b passes through the origin, yielding a 2% increase in cancer cases at dose level 20 (the intersection of a line drawn upwards from point 20 on the horizontal axis, and the curve line b). This kind of linear straight curve, or linear extrapolation, has always been used for genetic damage. Later, this linear relationship was also used for cancer risk. (BEIR I, 1972). This curve corresponds to a constant effect per unit dose from a high dose level down to a zero dose level.

c. This quadratic curve c corresponds to the hypothesis (or scientific assumption) that assumes a rising risk per unit dose level, and thus, yields a negligible effect at very low doses, for instance, for cancer and leukemia.

d. This corresponds to a curve with a true level of tolerance as was observed, for example, in the case of immediate death from high full-body doses.

a. This "supralinear" curve, which arches upward, will be discussed later on in relation to the Petkau Effect.

Does the Linear Curve Represent a Conservative, Safe Estimate?

If we now move up the vertical line from dose 20 to ascertain the risks at the various curves, we find an estimate for zero cancer cases using curve d, an approximately 0.2% increase in cancer cases using curve c, and a 2% increase in cancer cases with curve b. In other words, we get a greater or lesser risk for small doses of radiation, depending on the curve used. Leaving curve a aside (we will get to it later), curve b, the linear extrapolation, yields the greatest risk. It has therefore been regarded as the most conservative estimate, and seems appropriate for the requirements of radiation-protection. The famous BEIR I Report itself, was based upon this estimate. According to K. Z. Morgan, the risk of cancer at 1 rem differs between curves b and c by a factor of 100.[120]

For genetic damage, this linear relationship (curve b) between dose and effect had always been assumed. This relationship had been confirmed, too, by experiments with animals. In the case of cancer, however, the curve c tolerance levels was accepted only until 1960. Since then, that risk too has been calculated according to curve b. This calculation seemed to be on the safe side, since other curves (c and d) indicating a lesser risk were thus also covered. After all, radiation-protection was supposed to conservatively assume the highest risk consistent with available information, so as not to underestimate the danger.

The Japanese Atomic Bomb Victims

Except for the Japanese atomic bomb victims, there is hardly another test group available from which dose-effect curves may be drawn. It would not be unreasonable to expect to obtain important data from the X-ray and nuclear branches of medicine—doctors, nurses and patients. Apparently these groups are too small, the dose ranges too narrow and the dose levels too low to permit such studies. In any case, the study of survivors of Hiroshima and Nagasaki continues to be the best source; in fact, it is virtually the "Bible" of radiation-protection.[67] However, it has serious drawbacks:

a. The dose assigned to victims is not exact, as UNSCEAR points out:

> Despite continued investigations, the doses received can only be inferred from the distance of the survivors from the hypocenter of the bombs. The doses therefore are highly uncertain, and this uncertainty reflects on the dose-effect relationship.[194]

b. The tissue sensitivity of the irradiated Japanese may not be applicable to the general population. The UNSCEAR remarks in this regard:

> The surviving population has been heavily selected by the lethal effect of the irradiation itself, so that the survivors may not necessarily be representative of the irradiated population with respect to sensitivity to radiation carcinogenesis.[197]

And the ICRP adds:

> Even the Japanese population was not wholly 'normal': there was a deficiency of fit adult males absent on war service.[76]

c. In addition, the ICRP and the UNSCEAR point out that the Japanese were not uniformly irradiated. Some victims were in houses, behind walls, etc.:[196] "One-third to two-thirds of survivors who received a nominal dose of 650 rads and more were heavily shielded."[72]

d. The fact that the study of survivors did not begin until 1950, five years after the explosion, is cause for misgivings.[179] As a result of the catastrophic effects of the bombing,* the weakest individuals already died. Had they lived, many more might then have died as a result of the radiation exposure they suffered. Current cancer statistics do not take this into account.

This illustrates just a few of the uncertainties upon which our radiation-protection legislation is based.

The Risks of Radiation Keep Growing

It is becoming clear that we have reason for increasing concern. Although the commissions which set the tone for radiation-protection, the ICRP, UNSCEAR and BEIR commissions, periodically publish their progress reports, the dangers of the table below demonstrates how science is constantly betraying itself.

*In the initial period after a catastrophe, food, water, medicines and shelter are lacking, and unsanitary conditions prevail. Thus, those who are of poor health, or are lower on the social scale, generally die in disproportionately large numbers.

Data on somatic radiation risks of average adults:
Explanatory Notes:

1. Mortality: Expected radiation-induced leukemia and deaths from cancer per million persons irradiated with a whole-body dose of 1 rem (10 mSv).
2. Incidence: In contrast to mortality, this refers to all persons contracting leukemia or cancer, regardless of whether they die or survive.
3. In 1958, the ICRP did not yet believe that there was any risk of leukemia or cancer at low radiation doses. The dose limits for the general population were set primarily for genetic reasons.

MORTALITY: (Deaths from Leukemia and Cancer)[1]

Reference:	Year:	Deaths from Cancer/ Leukemia:	Remarks:
[a]ICRP	1958	0	(see note 3 above)
[b]ICRP	1966	20	Only leukemia
[c]ICRP	1972	40	20 leukemia, 20 cancer (+ 20 non-fatal cases of thyroid cancer)
[d]BEIR-I	1972	50 - 165	Leukemia and cancer
[e]ICRP	1977	100 - 125	Leukemia and cancer
[f]UNSCEAR	1977	75 - 175	Leukemia and cancer
[g]UNSCEAR	1982	100	Leukemia and cancer
[h]BEIR III	1980	10 - 501	Leukemia and cancer

INCIDENCE: (Leukemia and Cancer Cases)[2]

[i]BEIR III	1980	260 - 880	among men
[i]BEIR III	1980	550 - 1620	among women

---

[a]BEIR-I 1972, pp. 42, 60.
[b]ICRP # 9, item 95, pp. 16-17
[c]ICRP # 8, p. 56 Table 15.
[d]BEIR-I 1972, pp. 167-168; 91.
[e]ICRP # 27, item 38/39, pp. 13-14; item 33, pp. 12; # 26, item 60, p. 12.
[f]UNSCEAR 1977, p. 414, item 318.
[g]UNSCEAR 1982, p. 11.
[h]BEIR-III 1980, p. 145.
[i]BEIR-III 1972, p. 246.

The table shows that by 1980, radiation was considered more than 1000 times more dangerous than was assumed in 1958. Nonetheless, the ICRP has not reduced the maximum permissible dose for occupationally-exposed personnel, individuals in the general population, or the population as a whole.

In addition, the risk calculations of the ICRP and the UNSCEAR take only deaths into consideration. However, it is now known that thyroid cancer and breast cancer in women are the most common radiation-induced cancers, and for these two sites, mortality data give an inadequate indication of risk. This inclusion of data would ensure that all cancer victims—both the dead and those who recover—would be incorporated into the calculations.[28] Although thyroid and breast cancer can often be cured, they result in major psychological, social and economic costs to the affected, predominantly women endangered by breast cancer. According to BEIR III 1980 (see table), cancer incidence in women is in the range of 550-1620, as opposed to 260-800 in men (assuming, in all cases, 1 rad of one-time irradiation of 1 million men or women). Thus, women have the most reason to oppose nuclear power.

The ICRP and UNSCEAR hide the fact that a woman cured by the removal of a breast, or a patient recovered from thyroid cancer have also fallen victim to radiation. The sufferings of these people are callously disregarded, although many will require medical care for the rest of their lives. Cynical whitewashing of the risks of radiation is designed to ensure the continued existence of the nuclear industry. The official 1980 BEIR III report notes:

> Many members of the Committee believe that the incidence of radiation-induced cancer provides a more complete picture of the total social costs than does mortality [deaths].[26]

Moreover, it is assumed that the "incidence level"—the total number of cases of radiation-induced cancer—is approximately twice as high as the number of deaths from such cancer.[18,88] What the table then shows, is that ever greater uncertainty is reflected in the data with the passage of time; thus, the 1980 BEIR III report gives a range of uncertainty of 10-501 deaths from leukemia and cancer. This is partly due to the fact that the true risk of cancer is not known even for high dose levels. The observation period of the Japanese victims had been too short to determine this risk. Only after the population group has died off—which could take decades—or the incidence of radiation-induced cancer begins to drop, will a clear calculation become possible, and so-called relative and absolute risks give the same results.[29]

In addition, the 1980 BEIR III report also states:

> There is evidence that the increased risk of leukemia and bone cancer does not persist indefinitely but becomes negligible 25-30 yrs after the end of irradiation. For all the other radiation-induced cancers . . . the minimal latency period is 10 yrs. or more and there is as yet no indication that the increased risk of cancer eventually declines. There are however no epidemiological studies in which follow-up was carried out to the end of the life for the entire population cohort. Hence any projection of risk over the lifetime of the exposed persons involves considerable uncertainty.[27]

Currently, the cancer incidence rate among the Japanese is in fact increasing strongly. No one knows whether this will continue, and if so, for how long.[3, 50, 99]

Intrigues and Manipulation of Science (BEIR III 1980)

Scientists are of different minds as to which does-effect curve to choose for risk estimation of small radiation doses. This is an added reason for the great range of uncertainty in the data. What is ominous, however, is that this dispute seems to be less an academic debate than a hard-nosed economic and military tug-of-war. This, at least, is the implication made by Rotblat in his article in the *Bulletin of Atomic Scientists* of June/July 1981:

> Risk estimates for low-level doses may differ by as much as two orders of magnitude depending on whose theory is applied and which observed data are used. Unfortunately this is not merely of academic interest. The selection of a model for risk calculation can bring with it enormous cost differences. Powerful outside interests are involved. . . . The best illustration of this is the BEIR III report.[145]

The BEIR committee is a consultative committee of the National Research Council (NCR) of the United States National Academy of Sciences. Its first report, BEIR I, 1972, became a widely accepted reference work for radiation-protection of the population. The constant debate over atomic energy, and the need to continually analyze more current research results on low-level radiation led the Academy in 1976 to request a new report from the BEIR Committee. Under the chairmanship of E. P. Radford, professor of Environmental Epidemiology at the University of Pittsburgh and chairman of the BEIR's subcommittee on somatic damage, the committee completed its work at the end of 1978. According to Radford, no further meetings of the committees took place thereafter. The BEIR III report was then published and distributed in May 1979.

Then something very unusual happened. The Academy withdrew the report, though 17 of the 22 members of the committee had approved it. Moreover, the five dissenting members, including H. Rossi, had already been given ample opportunity to express their opinion. The Academy appointed a group of seven committee members to revise the chapter on cancer risk. A year later, the final BEIR III report of 1980 appeared. The main modification was that instead of a linear dose effect, a linear-quadratic curve was adopted. The latter curve differs particularly in regard to the predicted effect of small doses. Evidently this data was in accordance with orders. It showed a much lower risk of cancer between the linear and quadratic curves in the above graph of dose versus effect. Unsatisfied, Rossi argued for a quadratic curve with an even smaller risk result. In the final form of the 1980 BEIR III, both Radford and Rossi were able to state their different opinions.[31]

Deaths from Cancer:

Linear curve b (Radford):	167-501
Linear-quadratic curve: (Between curves b & c; the "official" curve)	77-226
Quadratic curve c (Rossi):	10-28

In his statement in BEIR III, 1980, Radford writes:

The new version of the report ignores all studies of cancer risk in humans with the exception of the Japanese data. It also reduces estimates of risk calculation to the extent that they approximate those of the BEIR I Report of 1972. What is not taken into account here is that an important step is missing namely that cancer risk should henceforth be described in terms of incidence; also much data is ignored which indicates that the risk of cancer is increasing, that the effective doses are continually dropping, and that the various types of human cancer that can be caused by radiation have increased. . . . The new version of the report has ignored these important points and therefore does not correspond to the latest state of knowledge regarding the risk estimation; this should in fact have been the task of the BEIR III Committee.[31]

In the same report, Radford illustrates the risk of cancer as follows:

Lifetime Risk of Cancer (Incidence) in 1 Million Men and Women[30]

	Men	Women
No irradiation	283,000	285,000
Additional cases, with life-time continuous irradiation of 1 rad/year	16,200-31,100	37,600-185,200
Additional cases, with a one-time irradiation of 1 rad/year	260-880	550-1620

Radiation-protection hereby becomes a pawn of interest groups whose concern for the continued existence of humanity is either minuscule or hypocritical. This is frightening, for according to what we know today, there should not even be any balancing of economic factors and risks. The utilization of radiation and radioactivity should be strictly limited to the realm of medicine. Technology based on nuclear fission should be banned immediately. In 1958, it was believed that doses of 1 rad/year would mean no risk of cancer. At that time, it was even unknown that women were more vulnerable to radiation than men, mainly due to cancer of the breast.

Nuclear Safety?

Radford notes, in regard to the BEIR III data, that radiation-induced cancer could not be ascertained in all cases, in spite of the unrealistically high radiation exposure rates. Thus, a situation amounting to radiation-generated mass slaughter could occur, while nuclear propaganda continued to claim that no one had ever died as the result of radiation.

To demonstrate the validity of this charge, we need look no further than a statement made to the Swiss newspaper *Basler Zeitung* of May 13, 1981 by Dr. H. P. Hänni, a physicist at the Swiss Federal Institute of Technology and nuclear safety expert at the nuclear power plant in Beznau, Switzerland: "Numerous investigations have shown that an annual radiation dose of 2 rems or 2,000 mrems (which corresponds to 60 rems in 30 years) cause no demonstrable damage. It is not entirely negligible, but the effect is so slight that no discernible damage can be detected. Other effects are dominant." Yet—according to official data—irradiation of a million people with 60 rems (0.6 Sv) over a period of 30 years would cause mass slaughter (see the above table). How could such a harmless picture of an accumulated dose of 60 rems (0.6 Sv) be painted as late as 1981¿ Must we citizens accept this as "nuclear safety¿"

Forced Medical Treatment of the Public with Artificial Radiation

The above risk calculations demonstrate all too clearly how the authorities protect and propagate nuclear power in the most criminally negligent manner. Even the mighty chemical industry seems an insignificant dwarf by comparison. Consider a recent Swiss-based scandal:

In early 1984, certain medications for rheumatism caused a great public uproar. An internal memo from the Ciba-Geigy Corporation had been leaked, revealing that the use of butazolidin and tanderil would lead to 1182 deaths within 30 years, in a patient population of 200 million.[127] Some critics were able to calculate estimates of deaths as high as 11,000 among 180 million people treated.[127] On the basis of one million affected persons, this would mean between six and 61 deaths. A high death rate as the result of medication, used only in a targeted, individual and voluntary manner, and which, most significantly, would not affect healthy individuals, was enough to cause the West German Federal Health Agency to issue a restriction. The use of medications containing phenyl butazone and oxyphenobutazone was limited to cases of Bechterew's disease (chronic inflammation of the joints, especially in the spine), and attacks of gout—and this only during a seven-day treatment.[5, 127] Additionally, in 1985, Ciba-Geigy halted sales of all internally-administered forms of tanderil worldwide.[130] Still, what about the products of nuclear fission? Artificially-generated radioactive substances mainly effect living things internally.

Compare the above situation with the lifetime risk of cancer (incidence) from only a single irradiation with 100 mrem (1 mSv) per year. This is the dose with which, according to the ICRP in 1984, individuals in the population can be repeatedly irradiated over an extended period.[90] According to the linear relationship in BEIR III, 1980 (Radford), we can assume 26-88 cancer cases per million men and 55-162 per million women. This does not even take into account the possible genetic damage and additional somatic effects of radiation other than cancer. According to Swiss Guideline R-11, the population in the vicinity of a nuclear power plant is permitted to receive an exposure of up to 20 mrem per year![103] Based on the experience with rheumatism medication, if radioactivity were a medication it could not be placed on the market, let alone be prescribed on a continuous basis. However, in order to implement nuclear power, such high risks are legally permitted. Moreover, persons exposed to radiation on the job are legally permitted much higher radiation doses than the 100 mrem/year (1mSv) being considered here.

It is unconscionable to continue to expose our environment and the entire population (not only sick people have to be medicated) to artificially radioactive substances. Because we naturally have some radionuclides in our bodies is no reason to add more. On the contrary: now that the danger of radiation is recognized and known to officialdom, any increase should be avoided. No one knows when these artificially-produced radioactive substances will show up in the air, water, or our food. They can be passed on from one living thing to the next. We should not allow ourselves to be deceived by low doses, for these poisons accumulate in our bodies. The existence of natural radioactivity should no longer be tolerated as an excuse.

So who has the right to establish "maximum permissible dose levels" for entire populations? Even according to the ICRP, biological damage begins at point zero. The Commission recognizes the assumption that there is no such thing as a threshold for tumors or genetic damage.[73] Rather, effects are qualitatively dependent on the probability of harm per rem and on total dose throughout the entire range of exposure starting at zero.

The ALARA Principle Trick

The ICRP's device for protecting the nuclear industry is to establish or maintain high dose-limits while at the same time applying the so-called ALARA principle—an acronym for "As Low As Reasonably Achievable," including the following adjunct: "Economic and social considerations [should be] taken into account."[81]

Of course "social considerations" should include ethical ones as well.[79] The cost-benefit analyses recommended by the ICRP stands in contradiction:

> The acceptability of levels of exposure to radiation proposed for a given activity should be determined by a process of cost-benefit analysis.[82]

This has not prevented the ICRP from establishing an excessively high maximum dose of 170 mrem (1.7 mSv) per year (over 30 years) for the world population, or from adding that this value need not represent true cost-benefit balance. What happened to ethical responsibility?

Inhuman Calculations

Demonstrating its inconsistency, the ICRP has also expressed risk estimates in monetary terms:

> The Commission discussed the use of risk estimates to assess the actual number of cases of disease that may be caused by any given exposure of individuals or populations. . . . One way to improve

the usefulness of risk estimates is to convert them into estimates of detriment expressed in monetary terms.[83]

Various authors have already conducted such inhuman calculations, and have placed a price-tag of "$10 - $250" on each "man-rem." Here is the callous formula used by the U.S. National Academy of Sciences to translate human suffering into dollars: Annual costs of illness in the U.S.A. were estimated at $400 per capita overall in 1970, based on a total bill of $80 billion for a population of 200 million. Lederberg then assumed that the permissible 170 mrem/year (or 1.7 mSv), which comes to 5 rem (50 mSv) over 30 years would raise the total illness level of the United States by between 0.5% and 5%. Thus, one rem would cause an increase of between 0.1 and one percent. Over the course of 30 years, medical costs per capita would normally be $12,000 (30 x $400), so that one rem (10 mSv) would cause additional medical costs of between $12 and $120 per person per generation (i.e. 0.1% - of $12,000).[16]

The standard is based upon medical costs and the loss of working days. Radiation victims cannot simply be trivialized according to the technological and economic logic of cost-benefit analyses. Health and human suffering are more than mere economic values and must not be reduced to a common denominator with economic values. Higher human and ethical values are destroyed by reduction to monetary terms—and the threat to life increases.

Calculations are particularly insidious when the inexorable occurrence of genetic damage is considered. In this case, the beneficiaries and the cost-carries are not even identical, for the latter are the future generations. The most blood-thirsty tyrants in history lacked this infamous opportunity—they could torture and murder "only" their contemporaries, but were unable to purposely saddle future generations with cripples, diseases and degeneration—to say nothing of the possible destruction of the ecology.

The Perfect Crime—Planned Legally?

In radiation-protection, ever more clever methods are being devised to establish maximum doses, in order to maintain the foundations for the existence of the nuclear industry. At an informative meeting of the Swiss Association for Atomic Energy (SVA) held on March 23, 1973, Professor

Dr. W. Jacobi, director of the Radiation-Protection Institute at Neuherberg (near Munich) and a member of the ICRP, delivered a paper that introduced an entirely new concept.[96] Significantly, the Swiss Radiation Monitoring Commission (KÜR) immediately accepted this concept without criticism.[100]

Jacobi introduced a so-called "maximum acceptable radiation risk" for the entire population (without asking the latter, of course), and assumed—amazingly—that a meaningful risk from radiation exists only if it is detectable. Thus, the risk should only be low in order that it be statistically indiscernible in the already-existing natural (spontaneous) incidence of cancer in the overall population. Hence, certain radiation risks may be considered acceptable and reasonable. Western Germany, by his calculations (based on 2350 spontaneous deaths from cancer per million inhabitants), should find a maximum of 10 deaths from radiation per year per million inhabitants acceptable. In a population of 60 million, this would mean 600 deaths per year from artificial radiation—which would not have to be recorded and could not be proven. How is such a thing compatible with the norms of a constitutional system? Human beings are being sacrificed, reduced to statistical figures as if they had no identity. The suffering is born by the victims and their families, with no compensation. The perpetrator has made sure of his own safety. Is this radiation-protection a contradiction in terms?

Jacobi adds that 600 deaths is a conservative number, since natural incidence of cancer is increasing anyway. According to the biologist Dieter Teufel, the logical consequence should be the state's tolerance of murder as long as it is performed with sufficient cleverness, and the numbers can not be statistically ascertained. "The assumptions underlying the model—that the additional risk of radiation to which an individual is subjected is all the more 'acceptable' the more other people also die of cancer (albeit of an utterly different cause)—is absurd on the face of it, as is the argument that the death of one or more humans becomes 'acceptable' by virtue of the fact that all other humans will ultimately die as well."[186]

Thus, Professor Gofman has called the use of radiation-protection legislation in the service of nuclear power a "license to kill." The minimal degree to which the ICRP showed concern for human life can be clearly seen in their 1973 report:

> However at the present time the relationship between dose and risk is not known with precision nor is it possible to make quantitative evaluations of the benefits. Despite this and in view of the continuing need for practical advice for planning purposes the Commission recognizes its responsibility to maintain its practice of recommending appropriate dose limitations.[80]

Even then, the ICRP should have established low maximum doses. Further planning of nuclear activities should have been halted—for one cannot "satisfy demand" by walking over dead bodies.

Nobel laureate Karl Lorenz stated, in regard to the untenable situation of radiation-protection: "What would a reactor safety expert do without his reactor? We are experiencing the curious phenomenon that the most vehement advocates of nuclear power are those who are supposed to be protecting us from it."[109] Some years ago, the chairman of the American Health Physics Society said: "The influence of our professional association grows with each new nuclear power plant . . . Let us put our mouth where our money is."

Even Radford (Chairman of the 1980 BEIR committee), in the June 19, 1981 issue of the respected journal *Science* stated that, by viewing radiation as ever more dangerous than had been assumed, he was swimming against the current of the general trend among radiation experts—that is, to minimize the dangers of radiation.[139]

Regrets for Radiation Workers: Five Rem Per Year Is too High

Where is the ethical social responsibility in the ICRP's maintenance of the same maximum permissible doses for occupationally-exposed personnel, though the risk of cancer since 1955 has increased by several orders of magnitude? If an employee contracts cancer he or she receives no compensation, as long as the acceptable dose has not been exceeded. No insurance company will recognize this illness as job-related—let alone pay for it.[144] The perpetrators are protected. By contrast, in any on-the-job or traffic accident the guilty party is ascertained and the victim, or his or her dependents, are compensated. If the ALARA principle were used for traffic, a speed limit of several hundred miles per hour would be permitted, with the recommendation to drive as slowly and as reasonably as possible.

Since 1978, experts have been demanding a reduction of doses for occupationally-exposed personnel, by factors ranging between 2 and 20—with no success.[120, 144] A reduction by more than a factor of 2 would present the nuclear industry with serious difficulties, since repairs of nuclear power plants and reprocessing facilities would not only become more expensive, but might even be rendered impossible altogether.[120, 144] Even now, according to a United States Senate report, hundreds of welders have to be recruited in order to reduce the danger for individuals, and many receive maximum permissible doses nonetheless. Others, due to carelessness, are subjected to even higher doses than permitted.[183]

J. Rotblat, former chairman of the British Institute for Radiology and the British Hospital Physicists' Association, stated in the *Bulletin of Atomic Scientists* the following in regard to the excessive maximum permissible doses:

> If dose limits were to be decided by what a given industry could afford, we might as well abandon all attempts at estimating risk factors from actual data and leave it to the economy to set radiation levels. These would then differ from country to country and even from industry to industry. Once we move away from scientific criteria the situation becomes untenable. . . . Even the [ICRP] admits that exposure over long periods at an average annual rate more than 0.5 rem (5 mSv) would entail a hazard higher than is acceptable for a safe occupation.[144]

Thus, the five rems permitted today are ten times greater than the unacceptable dose.

Health Damage due to Fallout

General

Our existing knowledge of radiation risk in humans is based on high doses applied over short periods, as in the case of Japanese A-bomb victims or irradiated patients. There is not yet any corresponding data for long-term doses such as occur with natural radiation, fallout from A-bombs, or emissions from nuclear power plants. We are making do with the theoretical linear extrapolation from higher to lower doses, and believe that we are thus acting cautiously. It has also been assumed—irresponsibly—that in view of the low risk, no damage can be detectable by statistical means. The information gap is particularly great in regard to small quantities of radionuclides incorporated into the body. These radionuclides are constantly emitting small doses of radiation. Nuclear tests cause fission products to be distributed throughout the world. Rainfall carries them from the atmosphere to the soil and into the ground-water, and thus, eventually, through the food chain to all living things. The quantity of the products formed and their half-lives are not the only decisive factor in determining the hazards of these products; just as important are concentration processes in plants, animals and humans, the absorption capacity for related chemical elements by organisms, and the place and length of time of their incorporation into the specific plant and animal cells.

Artificially produced strontium 90 which can build up in the bones over long periods of time, is particularly dangerous. Its 29-year half-life is relatively long. Moreover, it does not distribute itself evenly throughout the skeleton; rather hot spots occur depending on calcium requirements. It is impossible to predict where they may occur. Cesium 137, too, is one of the most dangerous long-lived isotopes—it builds up in muscle tissue. Iodine isotopes with the atomic weights of 131, 132, 133 and 135 are also especially threatening, since they concentrate in the thyroid; this is particularly critical for the unborn.

We will not be discussing the multifaceted biological behavior of all the various fission products. Much is known, but much is also unknown, particularly in regard to their effect on plants. The danger of tritium and carbon 14 for instance, has been appreciated only recently. We are far from a final evaluation of their effects—particularly in regard to forest death, as will be discussed in detail later.

The worldwide radiation level to which plants, animals and humans are subjected was increased by the atomic bomb tests of the late '50s. Milk is particularly vulnerable to strontium 90 contamination. The strontium content in the bones of infants climbed significantly. This led to a doubling of the natural radiation dose in the growing bones and bone-marrow of babies.

Eskimos showed cesium concentrations ten to 40 times higher than normal. Since the reindeer they use for food accumulated much cesium in their muscular tissue. These animals had ingested large quantities of cesium 137 by eating lichen, which had absorbed the cesium directly from the atmosphere.

Contamination dropped off after the partial U.S. test moratorium, but increased again as the result of the resumption of Soviet tests in 1961. Strontium 90 in the bones and iodine 131 in the milk or in thyroids of sheep can serve as sensitive indicators of such nuclear tests. Strontium is ingested with the grass eaten by grazing animals, and is passed on again in their milk. Milk and grain are among the most basic food sources, and like milk, grains tend to absorb these fission products. In this case, strontium is absorbed into the husks of seeds.

The Nonsensical Comparison with Natural Radiation

During 1960 and 1961, important food sources in West Germany were contaminated to a considerable degree with strontium 90 and cesium 137 as the result of atomic bomb tests. Some parts of the population were receiving 70% of permissible long-term exposure in their daily diet.[99] The German Federal Ministry of Economic Affairs considered whether the sale of whole-wheat meal and black bread should be banned and the degree of grinding of wheat limited. Since the radionuclides accumulate in the vitamin-rich outer layer of the seeds, it would have in effect been necessary to remove these outer layers, so important to a healthy diet, and to use the less nutritious white flour.

In 1982, the UNSCEAR illustrated the collective radiation dose of humankind resulting from atmospheric nuclear weapons tests as follows:

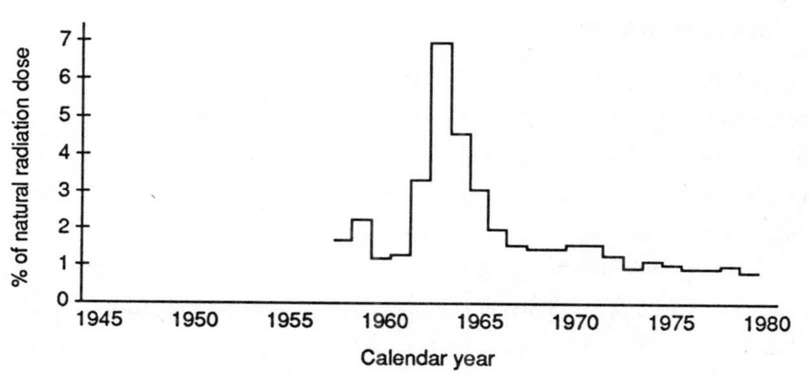

> There was a sharp increase of the annual collective dose in the early 1960s leading to a peak in 1963 corresponding to about 7% of the average exposure to natural sources.[201]

This is a good illustration of the nonsensical dose comparisons with natural radiation, which simply trivialize the danger of artificial radioactivity. Natural radioactivity does not lead to the same selective concentration in critical organs, nor does it lead to an increasing contamination of our environment.

No Official Studies

In effect, the U.S.A., Soviet and Chinese nuclear weapons tests amounted to a gigantic experiment with mass groups of people—in fact, with the entire population of the world. A large quantity of data on fallout and the accumulation of fission products in the environment was collected by means of a worldwide monitoring network.

This valuable data has still not been used in any epidemiological study by the governments of the world to shed light on the medical and genetic effects of an A-bomb fallout—and hence, the emissions of nuclear power plants. On the contrary, the information was in fact rejected. It was possible to continue claiming that fallout and nuclear facilities were harmless. This is a serious accusation which the scientific establishment, government organizations and, especially, the bodies responsible for radiation-protection—the ICRP, UN, BEIR etc.—cannot duck.

First Studies by Independent Researchers

Fortunately, independent researchers had different ideas. With modest means, they indirectly forced the nuclear test ban of 1963. A year earlier in *Science* magazine, the noted nuclear physicist, Ralph Lapp, had for the first time demanded epidemiological studies.[177] These studies would involve health damage in very large population groups.

Lapp was referring specifically to an A-bomb test conducted in Nevada on April 25, 1953. Its clouds had moved high over a number of states, and its fallout had come to earth in the course of unusually heavy rains in upstate New York, parts of Vermont and Massachusetts. Approximately half a million people were exposed to strongly radioactive rain. In the city of Troy, the radioactivity in the rain water was 270,000 microcuries, compared to an AEC maximum limit of 100 microcuries. This occasion would therefore have been a unique opportunity to undertake epidemiological studies, since unmistakable results could have been drawn.[177]

If a study had been allowed to show that many children had died due to the military test explosions in Nevada, the public would almost certainly have put up strong resistance to nuclear tests and gigantic peacetime nuclear energy program. Such studies would not, of course, be pursued by the AEC, a committee, after all, responsible not only for the safety but also for the promotion of nuclear power. This amounts to expecting the fox to guard the chicken-coop.

By that time, there were other scientists who were pointing to the danger of nuclear tests. One scientist was Professor Ernest J. Sternglass, radiology professor at the University of Pittsburgh. (Sternglass was also president of the Pittsburgh chapter of the Federation of American Scientists, a member of the American Physical Society, the Radiological Society of North America and the American Association of Physicists in Medicine. He had testified as a low-radiation expert before the Joint Committee on Atomic Energy and many other committees in the U.S.A. and abroad. Sternglass has written two books, *Low-Level Radiation* and *Secret Fallout*, which have attracted considerable attention.[165, 177] A large number of scientific studies of the biological effects of A-bomb fallout and radioactive emissions from nuclear facilities attest to his tireless research activity.) Sternglass was confronted with the problem of fallout for the first time in 1961. In the context of researching the feasibility of fallout shelters and evacuation, he came upon the tragic fact that the government assumed that adults could tolerate the enormous doses of 200 rads within a few days, and 1000 rads (10 Gy) over the period of a year.[177] No account whatsoever was documented of the resulting delayed damage in the survivors of their children. Only the effects of external radiation from fallout were considered.[177]

In those days, it was already known that radioactive iodine 131 accumulates heavily in milk and, when ingested, in the thyroid. As a result, the thyroid is much more strongly affected than if the iodine 131 in the fallout were only acting externally.[177] Scientists were also aware that unborn babies are many times more susceptible to radiation than adults. In addition, the doses in their thyroids were between 100 and 200 times as high as that in adults.[177] Also, research by Dr. Alice Stewart at Oxford University had already indicated that even one or two rads of X-ray exposure (10-20 mGy) could double the cancer risk of a child if it occurred in the last months of pregnancy.[112, 177] Just one tenth of that dose would have the same effect if administered during the first three months of life.[112, 177]

Sternglass concluded that the majority of children born after a nuclear war would die of leukemia or cancer, or would be deformed. Moreover, these hundreds of rads referred only to the external doses of the fallout, not to the contaminated foods.[177] Finally, after some resistance, Sternglass was able to publish an article in *Science* in the spring of 1963, causing considerable stir.[159] There, for the first time, he raised the question of whether a mother who was exposed to a fallout could not also harm her unborn child, as Stewart had already shown to be true for X-rays.

The 1963 Nuclear Test Ban

As early as 1958, Nobel laureate Linus Pauling had calculated that the fallout of the 1958 nuclear weapons tests alone would probably result in 15,000 children a year all over the world born with serious genetic deficiencies, 38,000 children would be stillborn, and 90,000 would die in the womb.[134] In the Spring of 1963, radioactivity in milk reached surprising levels in the U.S.A. These factors caused even more scientists, as well as the general public, to have serious misgivings. Though the AEC continued issuing reassurances, growing public pressure finally caused President John F. Kennedy, to propose the Atmospheric Test Ban Treaty with the Soviet Union and Britain in June of 1963.

In his speech to the nation in which he called for ratification of the test ban, Kennedy said:

> . . . the number of children and grandchildren with cancer in their bones with leukemia in their blood or with poison in their lungs might seem statistically small to some in comparison with natural health hazards but this is not a natural health hazard—and it is not a statistical issue. The loss of even one human life, or the malforma-

tion of even one baby—who may be born after we are gone—should be of concern to us all. Our children and grandchildren are not merely statistics toward which we can be indifferent.[177]

As moving as these words were, they had no educational effect on the governmental bureaucracy or on the AEC. The health agencies particularly showed themselves to be unusually committed advocates of the supporters of nuclear weapons.

Sternglass's Studies

Professor Sternglass undertook comprehensive studies on the effects of fallout in America and worldwide. He concluded, among other findings, that by 1968 in the U.S.A. alone, 400,000 babies under a year old had died of the consequences of radioactive fallout by 1968.

On the following pages, the curve shows the infant mortality rate in the U.S.A. between 1935 and 1980 for those under one year of age, due to pneumonia and influenza. The flattening of the curve may be considered the effect of nuclear weapons tests, especially since the normal decline resumed after the test ban of 1963.[173]

Sternglass can also base his conclusion on a major study in 1973 at Johns Hopkins University, in which observations of children born to mothers who were X-rayed during pregnancy were evaluated over a period of ten years. It showed that of the children X-rayed while in their mothers' wombs, one in a thousand developed cancer by the tenth year of life, and risked death due to infectious diseases of the respiratory and digestive systems. There was a total of 18.3 deaths per 1000 births for X-rayed children due to all causes combined, as compared with 9.8 for the non-irradiated control group of white children who received good health care. The earlier the irradiation had taken place during pregnancy, the greater the risk. These findings therefore indicate a very strong effect of radiation on the immune systems of children.[173]

Sternglass further determined from data obtained from the U.S.A. Vital Statistics that for every child who died during its first year, between five and ten were miscarried. This meant that the total number of developing infants who fell victim to the fallout of A-bomb explosions in the U.S.A. alone probably reached two or three million.

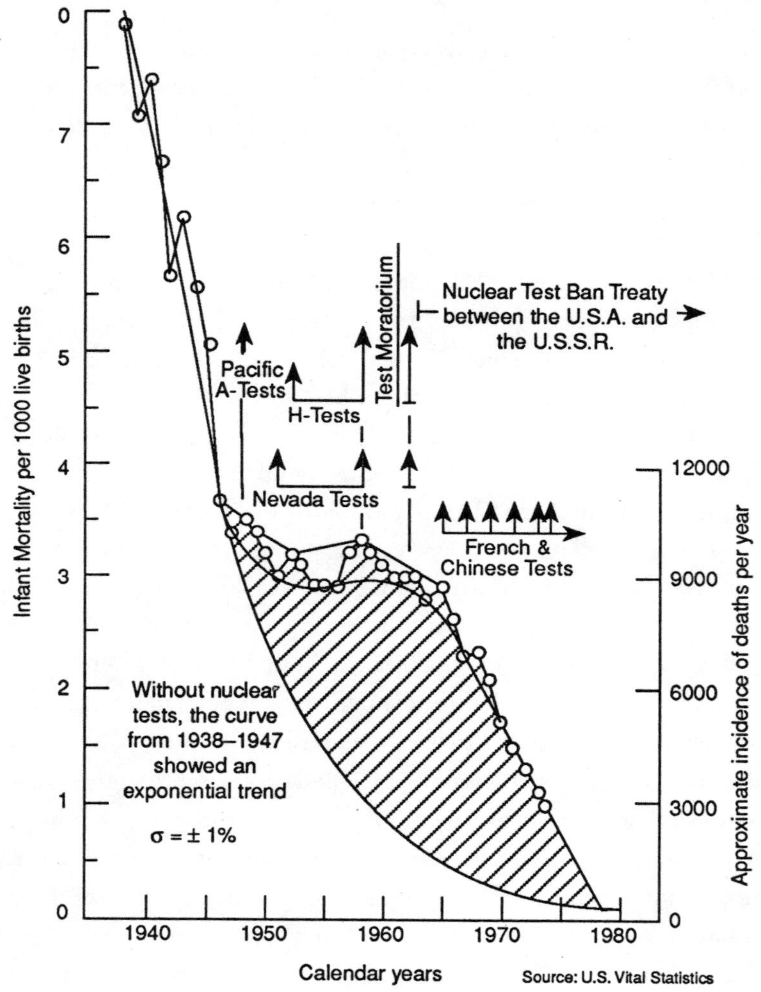

Sternglass reached analogous conclusions for many other countries, as well as in various areas of the U.S.A. itself.[162, 165] His findings were confirmed in the local effects of nuclear tests. However, in individual areas such as New Mexico, which due to climactic conditions had less rainfall or received less fallout, the data showed a steady fall in infant mortality (see the figure below).[162]

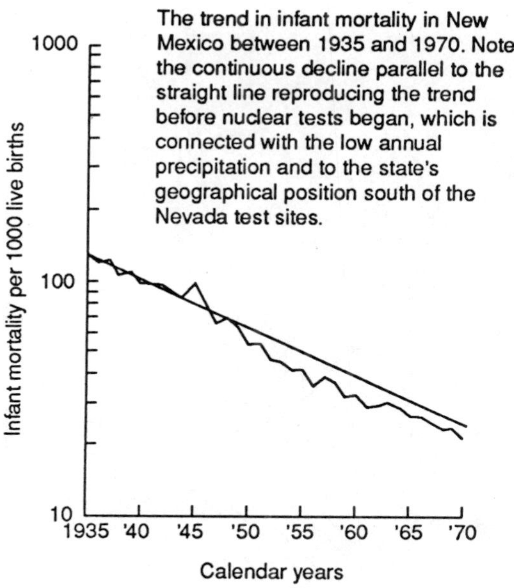

The trend in infant mortality in New Mexico between 1935 and 1970. Note the continuous decline parallel to the straight line reproducing the trend before nuclear tests began, which is connected with the low annual precipitation and to the state's geographical position south of the Nevada test sites.

On the basis of a 1972 study by Dr. M. Segi on behalf of the Japanese Cancer Society, Sternglass also found a steep rise in cancer mortality of up to 600% in five to nine-year-old Japanese children after the atomic bombings for all of Japan (see graph below).[152, 173]

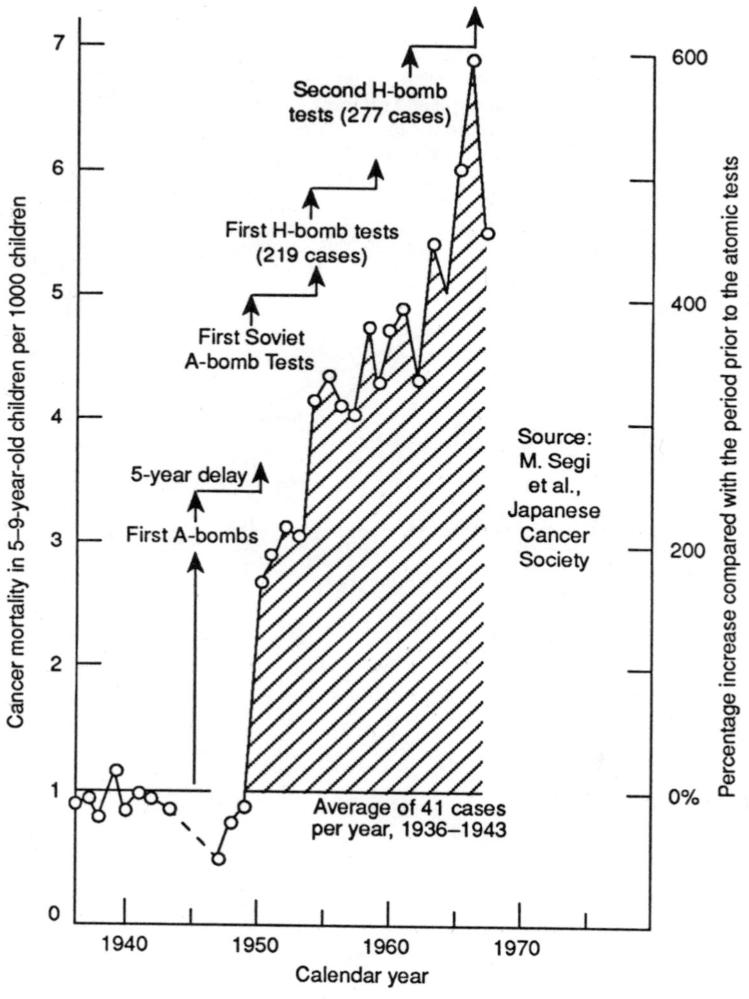

Study by L. B. Lave et al.

A study carried out at Carnegie-Mellon University in 1971, further supports Sternglass's findings.[105] Together with co-workers, the noted statistician Professor L. B. Lave investigated 61 urban areas in the U.S.A. between 1961 and 1967 using a multivariate statistical analysis. He found a high correlation between the death rates of all ages from a variety of illnesses, and the fallout level in milk. Lave recommended that his collection of more extensive and better data be initiated immediately. Neither government nor industry had any interest in doing so.

Study by Dr. C. E. Mehring

The first comprehensive study of the effects of fallout on the immune system was presented in 1971 by the German physician Dr. C. E. Mehring. At a nutrition congress in Montreux, Switzerland, Mehring shared his extensive statistical evaluations covering two periods of increased environmental radioactivity in Germany.[119] Due to nuclear bomb tests occurring during the 1950s and 1960s, the population indicated an increased susceptibility to common illnesses, a tendency for diseases to progress toward such complications as perforation of the appendix, and increased mortality from cancer, leukemia, and diseases of respiratory and other organs. The basis for the investigation was data from members of health-insurance funds, as well as the sickness and mortality statistics of the German Federal Armed Forces. Young people under 20 in particular, were seriously affected. There was also a reduction of immune defenses as measured by leucocyte depression. Thus, as early as 1971, Mehring hypothesized that low doses of radioactivity in the diet may have accelerated a wide variety of infectious and chronic disease.

An Unexpected Court Judgment

In 1984, for the first time in history, a United States court awarded damages to the victims of nuclear tests. Judge Bruce Jenkins of the federal District Court in Salt Lake City, Utah, found that the U.S. Atomic Energy Commission had behaved negligently during the 1950s and 1960s by not sufficiently protecting the inhabitants of three states from the consequences of atmospheric A-bomb tests.[128] In particular, the judge ruled that information given to the public on the known and predictable biological long-term effects of the tests had been inadequate. Furthermore, known methods available for preventing or reducing the effects of radiation had not been used. According to testimony from a series of witnesses, explosions were postponed when winds might potentially have carried the fallout to heavily populated regions, such as Las Vegas or southern California, while little or no consideration was given to the exposure of those inhabiting the rural areas just east of the test sites. Twenty-four individual suits were accepted for trial. Jenkins ruled that in ten cases, radioactivity was responsible for the resulting various types of cancer. A total of $2.6 million was awarded to the families of the victims, and to one survivor. It was a judicial case that established, for the first time, a connection between nuclear tests and cancer. The case was appealed to higher courts, and the judgment was reversed on technical grounds. The government was immune to damage claims. Hundreds of other claims from Nevada and nearby Utah and Arizona are still pending, but are now mute.[128]

In Britain, too, lawsuits were brought by British soldiers who participated in nuclear weapons tests near the Australian desert garrison of Maralinga. Of the 600 soldiers and civilians who were present in the test area, 114 have died—109, of cancer.

Damage to Health Due to Nuclear Power Plants

Studies by Independent Researchers

Until 1970, Sternglass, like most other scientists, believed that nuclear power plants under normal operation hold no danger. He then learned from government publications that some reactors, like the Dresden reactor in Illinois, were emitting as much as 260,000 curies, while others were emitting hundreds of thousands times less. The reason for this was that there are two types of reactors—boiling-water reactors and pressurized-water reactors. The boiling-water reactor was developed for submarines. Because radioactive gas-bubbles might betray the submarine's location, these reactors had to be as tightly secured as possible. Two separate coolant circuits were installed, and the pressurized-water reactor was born. For cost reasons, however, the industry was satisfied with only one cooling circuit. After all, they were competing with fossil-fuel fired plants. For this reason, they were willing to let the radioactive steam produced in the reactor circulate directly to the turbines. In fact, there is no such thing as a completely tight pump packing for the rotating bearings of turbines. Inevitably, numerous leaks developed. Sternglass thus expanded his statistics to include nuclear facilities with results analogous to fallout. Correlations were found with a variety of health risks: Increased infant mortality, often going hand in hand with a lower birth rate, increased premature births, increased death rates from cancer and leukemia, and deaths from arteriosclerotic heart diseases.

Correlations with increased infant mortality rates were found at the Dresden reactors, the Big Rock Point Nuclear Power Plant in Michigan, the Peach Bottom Nuclear Power Station in Pennsylvania, the reprocessing center in West Valley, N.Y., and the Indian Point Nuclear Power Station near New York City.[160, 161, 162, 163, 164]

The Dresden reactor was a classic case. Though Sternglass had to depend on relatively low numbers, the 140% increase in premature births in Gundy County, where the reactor is located, could not be ignored. Sternglass calculated that according to statistical laws, the likelihood of a chance occurrence was 1 : 10,000. In 1967, one year after the West Valley nuclear fuel reprocessing plant in Cattaraugus County, NY, went on line, the county's infant mortality rate rose to a level 54% higher than that for the State of New York as a whole. This rise could be observed also in all counties in the direct vicinity of this facility, and it fell off in proportion to the distance away from the plant. Toward the northeast, infant mortality declined to the level of the state average at a distance of some 50 miles away from the plant. But to the southwest, along the Allegheny River below Cattaraugus County, infant mortality dropped off slowly as far away as the Pittsburgh area. Even Armstrong County, 115 miles downriver, still showed a 4% increase in infant mortality. Evidently, according to Sternglass, it was not the inhalation of the airborne radioactivity that was decisive, but the ingestion of mild and drinking water. Gaseous fission products brought down by the rain, such as strontium, cesium and iodine were deposited on the land and entered the river.[165] Downstream from the area where the plant was located is where the river supplies drinking water.

The Cover-Up of the Shippingport Drama

In 1973, Sternglass uncovered serious environmental contamination by strontium 90, cesium 137 and iodine 131 at the Shippingport reactor (rated at only 90 Megawatts), which took place in 1971.[107, 166, 167, 168, 169] The strontium content of the soil in the vicinity of the reactor was one hundred times as high as anywhere else in the area, and fell off in proportion to distance from the reactor. At the same time, high strontium and iodine levels in milk were measured within a radius of ten miles of the reactor. The contamination could also be determined in the sediments of the nearby Ohio river, in food, and in the teeth of the local calves. Radiation exposure in the community of Shippingport was 180 mrads/year (1.8 mGy). Natural radiation there is 96 mrad. There were even months in which the level rose to 306 or 371 mrad/year. Sternglass was able to demonstrate increases in infant mortality and in the number of deaths from cancer. Consider that this was a "model" reactor, with the lowest emission rates—on paper, zero curies in 1971. This same year was later found to have been the year of greatest environmental contamination.

So much compromising documentation was available that for the first time in the history of nuclear power, a commission was appointed, by the governor of Pennsylvania, to determine whether or not the reactor had caused damage to health. In the hearings, Sternglass won support from such noted scientists as K. Z. Morgan, J. Bross, and M. Degroot.[6, 141] Still, the authorities and the power companies pulled out all stops in order to deny the incident.

In the final report of the investigative commission, presented to the governor in 1974, the authorities could circumvent the problem only by means of the most threadbare excuses, for instance: "The lack of any precise and comprehensive radiation monitoring outside the reactor during the previous operational years 1958-1971 makes confirmation of the published emission rates impossible. The high levels of strontium 90 in milk and of strontium and cesium in the food of inhabitants of Pittsburgh, as ascertained by government reports, is unexplained.[6, 141]

The importance of the strontium 90 measurement is shown by the statistics compiled by Sternglass shown below, according to which the strontium level in milk in the vicinity of the reactor varied directly with the power output of the reactor between January and June 1971. Characteristically, U.S.A. authorities stopped the reporting requirement for strontium levels after the accident near Harrisburg in 1979—for "budgetary" reasons.

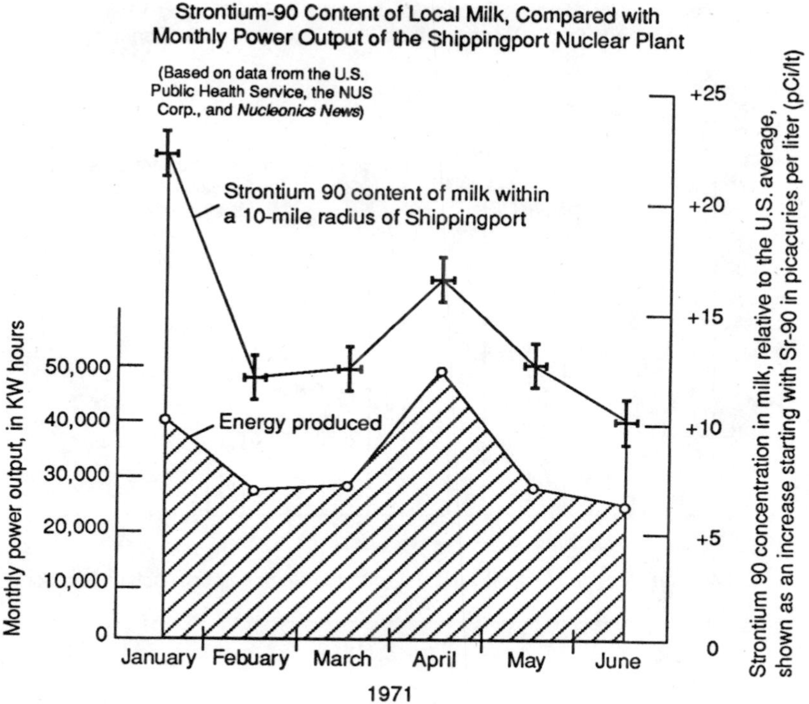

Over a ten-year observation period, Sternglass also found an increase in cancer mortality that declined with the distance from Shippingport (see graph below).[167]

Employees at the Shippingport reactor paid a high price. In 1979, Sternglass inspected 22 death certificates of reactor employees who had helped to clean contaminated pumps and other heavy equipment items. Of these 22, ten had died of leukemia or cancer—twice the normal rate.[177]

Research Reactors

Sternglass also uncovered deplorable conditions at the smaller Triga research reactors at various universities and research labs.[165, 177] Near the reactor at the University of Illinois at Champaign-Urbana, infant mortality rose by 300% between 1962 and 1965, when the reactor was brought on line. During the same period, the mortality rate due to congenital malfor-

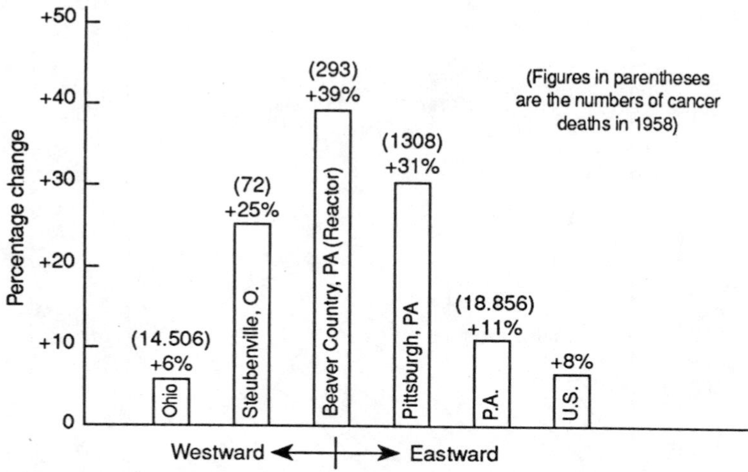

mations increased by 600%. When the reactor was shut down in 1968, the cases dropped again precipitously. The short-lived fission products reached the nearby population without significant decay, since the reactor was in the middle of an urban area.

Studies by Morris De Groot, Gerald Drake and John Tseng

Morris De Groot, Chairman of the Statistics Department of Carnegie-Mellon University, investigated the Dresden, Shippingport and Indian Point reactors, as well as that at the Brookhaven National Laboratories, in July 1971. De Groot confirmed the hypothesis of a connection between emission of pollutants and infant mortality, except in the case of Shippingport where later investigations revealed falsification of release data.[45]

Dr. Gerald Drake, a Michigan physician, published a study in 1973 on the community of Charlevoix County, Michigan, where the Big Rock Point reactor is located. He found the following data on health damage for county rates in excess of state averages, over ten years of plant operation:

Increased	infant mortality:	+49%
"	premature births:	+18%
"	deaths from leukemia:	+400%
"	deaths from cancer:	+15%
"	deformities:	+230%

Although all categories of damage showed increases, the results were based on small numbers. Drake recommended that more such studies be undertaken.[47]

A further important investigation was conducted by John C. Tseng at Northwestern University in Chicago, of seven nuclear facilities. All cases confirmed the possibility of a correlation between emission of radioactivity and infant mortality. In comparison, four coal-fired plants investigated could definitively be said not to yield any such conclusions. Tseng, at that time, made a very important recommendation: "Hitherto, whole-body irradiation has been viewed as most critical for environmental radioactivity. For background radiation, which is by and large gamma radiation, this may be a valid assumption. But for fallout and for the emissions of nuclear power plants which contain radiation, a specific organ of the body, or perhaps the embryo, might be a more appropriate critical organ."[193]

The Three Mile Island Accident near Harrisburg (TMI)

In the notorious TMI accident of March 28, 1979, a gigantic catastrophe was only narrowly avoided. The Catholic diocese had spontaneously empowered all priests to grant general absolution.[190] Some 170,000 people fled the endangered area. The reactor nearly suffered a total meltdown, and enormous quantities of radioactive gas had to be vented. When Sternglass flew to Harrisburg for a press conference 36 hours after the accident, his Geiger counter showed values from four to fifteen times higher natural background radiation. In the building where the press conference was held, readings of three or four times the normal level indicated the presence of significant amounts of radioactive gases.[177]

Thirty-six hours later, in Albany, New York, 225 miles away, the lab at the State Department of Health was still registering radioactive clouds carrying quantities of xenon 133 with between 3120 and 3530 picocuries per cubic meter (pCi/m^3) the normal mean level of this inert gas is

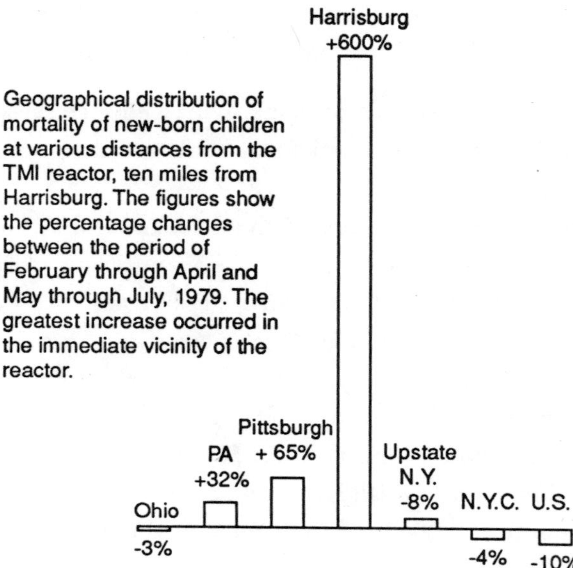

Geographical distribution of mortality of new-born children at various distances from the TMI reactor, ten miles from Harrisburg. The figures show the percentage changes between the period of February through April and May through July, 1979. The greatest increase occurred in the immediate vicinity of the reactor.

less than one thousandth of that: 2.6 pCi/m^3. This is not so surprising, considering that the reactor operator stated in a document labeled TDR-TMI-116, of July 31, 1979, that a quantity of radioactive inert gases equal to ten million curies had been emitted within the first six and a half days; of that, seven million were released during the first day and a half.[207] The quantity of iodine 131 emitted at the same time, was variously estimated at fourteen or between ten and twenty curies; again, the greater portion, as in the case of the xenon, had been released during the first 36 hours.[177, 190]

Sternglass points out that the evacuation of pregnant women, beginning only on the third day, occurred much too late. By that time the fetuses had already absorbed the iodine inhaled by their mothers into their thyroids.[177] The effect on the newborn infants is evident from his statistical chart reproduced below.[176, 189] The strong iodine emissions were also confirmed by the increased content of iodine 131 in the milk of grazing animals and the thyroids of wild animals in the vicinity.[49] L. von Middlesworth assumes that the slightly increased amount of iodine 131 (1-2 pCi/g) in sheep in Wales between the end of April and the end of May 1979 was due to the TMI accident, which happened 3600 miles away.[206]

Sternglass also stresses that the damage is not due primarily to the external full-body dose, but rather to a slight delay in thyroid function due to absorption of iodine from the fallout. This led to stunted growth, so that the fetuses were not yet completely ready for birth after nine months. Thus, the immature babies died in larger numbers from immaturity and respiratory defects. He was able to establish precisely that phenomenon in Harrisburg and Pittsburgh hospitals after the TMI accident.[176, 189]

Sternglass met with the objection that, even if a statistically-flawless correlation between low-level radioactivity and increased infant mortality and other health damage could be established, the causal relationship was nonetheless not proven.[189] He responded, justifiably, that by the same logic, one should stop opposing smoking, since there, too, the correlation was only statistically demonstrated, and no causal relationship between smoking and cancer could ever be provided. For this reason, no one has yet been in a position to claim damages from a cigarette manufacturer because he or she smoked two packs a day and developed cancer. The same is fundamentally true for all epidemiological investigations.

This argument is used worldwide by pro-nuclear governments and their advisors. For instance, in response to a Swiss parliamentary inquiry on statistical data gathering regarding causes of mortality, frequency of disease and other damage (leukemia, cancer, still-births, congenital defects) in the vicinity of nuclear power plants, the government responded:

> In the vicinity of nuclear power plants, the monitoring networks of the Radiation-Monitoring Commission (KUR and Energy Office overlap. The Federal Council* considers this type of supervision in the vicinity of nuclear power plants sufficient, and implementable at reasonable cost. . . . Moreover, the statistical significance of the difference between two mortality figures alone does not establish a relationship between cause and effect.[158]

*The seven-member Swiss chief executive body.

The biological, medical and possibly ecological effects of low-level radiation doses can be only insufficiently estimated today on the basis of purely chemical and physical measurements. Our concerns today are enormous, and they are now supported by the discovery of the Petkau Effect, a phenomenon which has until now been covered up. It is only by means of epidemiological investigations that we can learn what is really going on in the vicinity of nuclear power plants.

After all, such evaluations as that of cancer risks among A-bomb victims performed by the ICRP are based on epidemiological investigations. Thus, where such investigations fit the political bill, they are regarded as significant, and are widely respected; where this is not the case, they are deemed meaningless.

The TMI accident occurred when hundreds of cases were pending against the government by military veterans and residents of Nevada and Utah who blamed their ailments on the nuclear-weapons tests during the 1950s. In order to prevent a similar development after the TMI accident, the government and the operators of the plant denied that any damage had been cause.[177] Yet, a few months before the accident, a study by Dr. Joseph Lyon had appeared in the *New England Journal of Medicine* which showed that children living in the test areas during the A-bomb test period had a 2.5 times greater leukemia rate than in the preceding or following periods.[177] Moreover, a government study by the United States Public Health Service (the so-called Weiss Study), that had also found higher leukemia rates in these test areas, had been suppressed. A letter from the NRC to the Health Service was also made public, which states:

> Although we do not oppose developing further data in these areas [regarding leukemia and thyroid abnormalities] performance of such studies pose potential problems to the commission: adverse public reaction, lawsuits, and jeopardizing the programs at the Nevada test site.[177]

It is worth noting that following the accident, the then-Commissioner of Health of the state of Pennsylvania, Dr. Gordon MacLeod, admitted that there was a marked increase in infant mortality in a five-to-ten mile radius around the power plant. His successor, Dr. Arnold Mueller, announced that the infant mortality rate inside the ten mile zone was no different from that of the state of Pennsylvania as a whole.[111] MacLeod had been forced to resign after the accident. He has testified that since his resignation, a number of concerned employees of the Health Department called him up repeatedly to complain that abnormal health data were not being published.[111] As MacLeod noted, the official statement released in May 1980 is therefore incomprehensible. It had stated that after careful study of all available information, there had been no evidence of increased mortality among fetuses, infants or children caused by radiation from the nuclear power plant.[111]

Thus, as in the case of Shippingport, no health damage was officially acknowledged in Harrisburg. Later, a study published in 1987 by Dr. Jay M. Gould, head of the New York research firm Public Data Access, evaluating the data of the United States National Center for Health Statistics said that the age-adjusted U.S. mortality rate rose by a statistically significant 1.5% from 1979 to 1980. The greatest increase occurred in the states within a 500 mile radius of the Three Mile Island reactor. He estimates that some 50,000 "excess" deaths may have occurred in the U.S.A. between 1980 and 1982. Pennsylvania and New York, the closest states downwind from the reactor, had the greatest statistically significant increases in deaths, with 5.5% and 5.2% increases, respectively. Gould points out, cancer and genetic damage were the expected principal effects after radiation exposure. However, the new study raised the possibility that the most immediate and significant effect of low doses could be effects on the immune system.[218]

Study by Dr. Carl J. Johnson

Up to his recent untimely death, Dr. Carl J. Johnson was the chief medical officer at the South Dakota Department of Health in Pierre, and an associate clinical professor at the University of South Dakota Medical School. Johnson was a widely recognized authority in the field of radiation effects, and published many studies in which he found a correlation between emissions from uranium mines, fuel reprocessing centers, nuclear power plants, and fallout from nuclear explosions, with increased cancer risk for exposed workers and the nearby population.[221]

Possible Consequences of the Accident Still Unknown

Recently, familiar pseudo-scientific tricks have been used to trivialize the possible effects of nuclear accidents and catastrophes.[150] Pro-nuclear forces, such as the American Nuclear Society and other nuclear industry-funded organizations, have tried to prove by theoretical calculations that the accidental radioactive contamination of the environment would be much lower than has hitherto been assumed. They propose to simplify the process of issuing operating licenses and to relax the safety regulations.

The widely respected American Physical Society refuted the industry study by pointing out that it is impossible to assume that in a reactor accident only a small portion of the inventory of fissionable material would be released.[150] Chernobyl has since proven that the APS was right. Now, safety regulations are being tightened, and reactors are being "retrofitted." Once again, the nuclear establishment has been falsely optimistic.

Will Artificial Radioactivity Make Us Less Intelligent?

American psychologists have wondered for many years why the results of the routine Scholastic Aptitude Tests (SATs) taken by eighteen year-old Americans, have dropped off alarmingly since 1964; implicitly, IQs were dropping as well. In 1977, a report called the Wirtz Study was issued by a special commission appointed to investigate this matter. The commission processed some two dozen studies. The result was stunning. No single factor, or even any group of factors, could be clearly identified as responsible—including cultural or racial changes over the years, or more difficult tests. Earlier attempts at explanations, such as effects of the Vietnam War, increased TV-viewing, increased divorce rates, violence, crowded schools, lower-quality instruction, stiffer requirements, were all rejected.

By coincidence, Sternglass read in *The New York Times* that in 1975, SAT results had experienced their sharpest drop in twenty years—not two or three points, but ten. It struck him like a bolt of lightning: "When were these young people born, when were they in their mothers' wombs?"[177] Most of the young people were eighteen years old, and so had been born in 1957. This is the year of the highest measured fallout, due to the highest number of kilotons of A-bombs ever set off in Nevada. Sternglass was glad that during that time he had urged his wife and friends to give the children only powdered milk, so that the iodine 131 (half-life: 8 days) would have had time to decay.

Sternglass also recalled the Hanford Symposium of 1969. There, reports were delivered on the consequences of the A-bomb explosion, "Bravo", on the Marshall Islands. This fallout affected Rongelap Island, 152 miles away. Over the following 15 years, all the island children developed thyroid diseases, and showed retardation in both their physical and mental development.[154]

In an exhaustive study, in cooperation with the school psychologist, Steven Bell, at Barry College in Mount Barry, Georgia, Sternglass was able to demonstrate the connection between the average test-score decline of American young people and the fallout from A-bomb explosions during the 1950s and 1960s, mainly in Nevada and New Mexico.[175, 177, 178]

The graph below shows the close relationship between the declining SAT test scores between 1958 and 1982, and the beginning and end of

Verbal SAT scores in the U.S. for the years 1958 through 1982, and their relationship to the nuclear weapons tests at the times of birth of the testees, 18 years earlier,

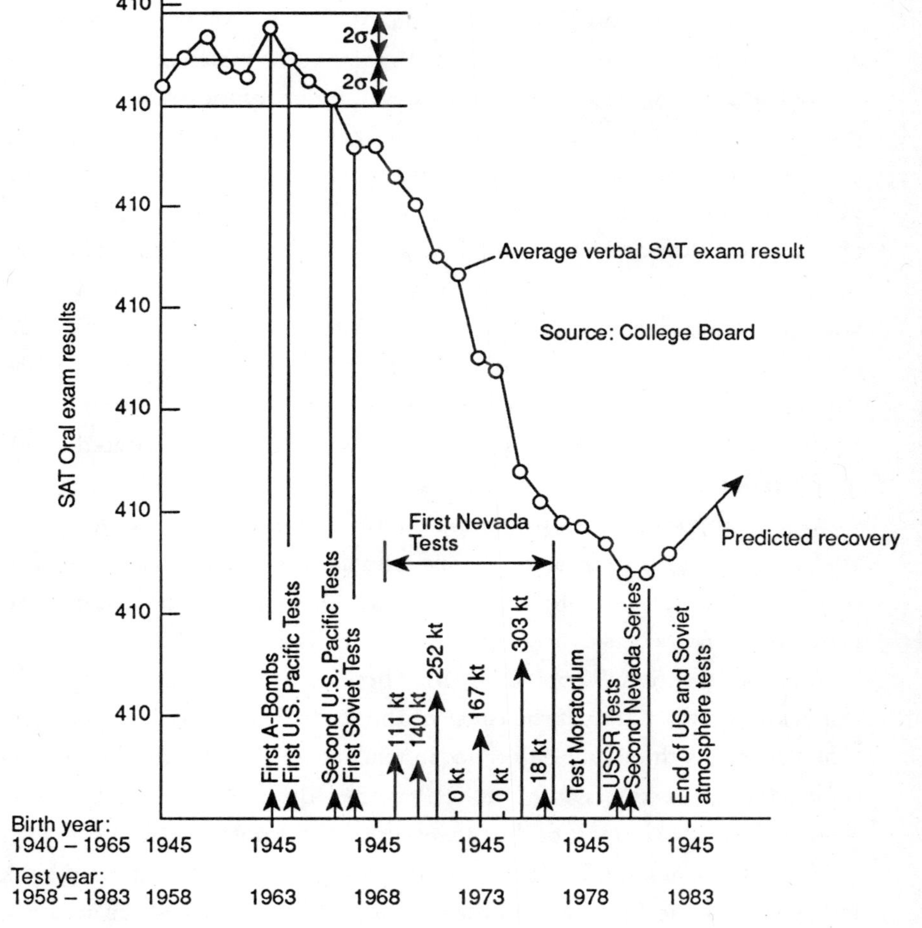

the nuclear explosions, particularly in Nevada, seventeen to eighteen years earlier. Since the results for the verbal and math tests show parallel curves, only the former is shown. Since no data is available prior to the late 1950s for radioactivity in milk, the kilotonnage of A-bombs exploded is used. Significantly, the test results have begun to improve since the test ban (the curve starts to rise again), just as Sternglass and Bell had predicted in their first publication on the matter, in 1979.

Since the 1960s, the U.S. Department of Health and Human Services and the EPA have also been publishing the concentration of the more significant radioactive fission products found in the pasteurized milk in the 50 states—iodine 131, cesium 137, strontium 89 and 90 and barium 140. [177]

Here, too, statistical connections with the SAT results could be found. The strongest correlation was with iodine 131, which builds up in the thyroid, and which in turn controls the development of the brain. There was much less correlation to the more long-lived cesium 137, which accumulates mainly in the muscles. No correlation was found with long-lived strontium 90 or short-lived strontium 89 which, like barium 140, goes to the bones—and may cause cancer of the bone, as well as damage to the immune system.

Comparing SAT tests by state during the years 1974 and 1976, Utah, the state closest to the Nevada test site, had the highest drop—26 points—as opposed to only two points for distant Ohio. The decline applies to the age cohorts born in 1956 through 1958; at the time, Utah had the highest iodine content in its milk, while Ohio was located south of the path taken by the fallout clouds.

Utah, with its Mormon population, normally reported very high SAT test results, in spite of its serious air pollution problems attributed to copper smelters and coal-fired power plants. Of course, the people of Utah have just as many cars as anyone else in the U.S.A.—but all these polluters produce no strontium 90 or iodine 131. Therefore, the factors causing traditional air pollution are not the cause of the sharp drop in comparison with Ohio. Nor was there any difference in teacher quality, TV use, or general socioeconomic factors. Mormons don't smoke, drink alcohol or take drugs. Nonetheless, the sharpest decline in test scores was noted in Utah.[177]

Sternglass and Bell also cite two studies, conducted at New York University Medical Center and the Chain Sheba Medical Center in Israel, reporting on the effect of radiation on mental function.[177] Two thousand two hundred and fifteen and 10,842 children, respectively, were treated for a fungus disease of the scalp (tinea capitatis) in the manner that was popular at that time—by radiation. During the following period of 20-25 years and 30 years respectively, there were not only greater rates of thyroid and brain tumors, as compared to a non-irradiated control group, but also mental problems which were evidenced in poorer performance in school and increased cases of psychiatric disorders. The thyroid dose was only 6–9 rads in these subjects—much less than the 10–60 rads to which the Utah children had been subjected.[177]

Sternglass and Bell discussed other details as well; all supported their hypothesis that fallout must in fact be the explanation for the extraordinary observations. B. Rimland and G. Larson, two psychologists at a naval personnel research center in San Diego, reached similar conclusions. They pointed out that the physical and chemical changes in the environment had so far been given far too little attention. They investigated the noticeably decreasing performance of young servicemen, and cited the Sternglass-Bell findings as a likely contributing factor.[142]

The ominous effect of radioactive iodine 131 on the thyroid's function, and hence on the mental development of unborn children, must constitute among the most devastating results of exposure to nuclear fission fallout.[177] The mere possibility that fallout from a few small nuclear weapons in Nevada may have resulted in a whole generation of mentally-impaired young people should be cause for alarm. In case of a limited nuclear war, both attackers and defenders would have to assume unforeseeable consequences for generations to come—merely as a result of the fallout which would contaminate the air, water and food. This would be true even if not a single city were attacked and not a single human being killed.[177]

Nuclear power plants, too, produce large amounts of iodine 131. Therefore, we must ask ourselves whether they, too—in normal operation or in the case of accidents like Three Mile Island and Chernobyl—can affect the ability of children to learn. Remember, iodine 131 is one of the normal emissions of nuclear power plants.[104] Significantly, an ICRP representative named Thorne, in 1986, was forced to admit that there is no safe threshold level for brain damage and mental retardation in utero. It would be ironic and tragic if a highly technological society were to see its intelligence reduced, when particularly such a society requires it most.

The Petkau Effect

A New Dimension of Radiation Danger?

In 1972, the scientist Abram Petkau at the Canadian Atomic Energy Commission's Whiteshell Nuclear Research Establishment in Manitoba made an accidental discovery deserving of the Nobel Prize.[135, 136] Petkau irradiated artificial cell membranes under water, using phospholid membranes which are similar to the cell membranes in living cells. He discovered that if the irradiation continued over an extended period, the membranes would tear after a much lower total absorption of radiation dose than if this total dose were emitted in the form of a short burst, as used for X-ray film.

A living cell consists of a cell membrane and a cell nucleus (see figure). But the cell membrance is not only there to hold the watery cell plasma together; it has many important functions in biological processes. These tasks have been compared with those of an entire industrial corporation. Thus, intact cell membrances are essential for a healthy life.

What Petkau then discovered was the following:

A *short-term* irraditaion of 26 rads per minute (i.e., a high dose rate)* from a large X-ray machine required the high total dose of 3,500 rads in order to destroy the cell membrance. However, with protracted radiation of only 0.001 rads per minute (i.e., a low dose rate) from radioactive table-salt ($Na^{22}Cl$) dissolved in water, a total dose of only 0.7 rads was required to break it.

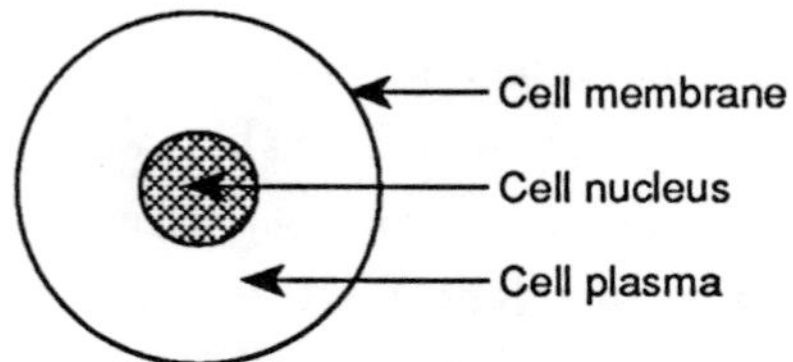

**Dose rate:* One might drink a large mug of beer in a brief period, which would correspond to a high dose rate. However, it would also be possible to drink the same amount of beer slowly, in a drawn-out manner, which would correspond to a low dose rate. The radiation dose would thus correspond to a certain quantity of energy absorbed by the irradiated body (rads). Thus, *a given dose of radiation* (or energy quantity) may be transmitted within a brief period, i.e. at a high dose rate, or else slowly and more drawn out, over a longer period. The latter would amount to a low dose rate.

Thus, in the case of low-level, protracted irradiation, a 5000-times smaller total dose was necessary (3500/0.7 = 5000). This was truly an incredible discovery.

Repeating the experiments numerous times, Petkau always reached the same conclusion. The more drawn-out the radiation, the lower the total dose required to break the membrane. This showed that small, chronic radiation doses could be much more dangerous in their effect than high, short-term doses. This revolutionary new discovery is in sharp contrast to the genetic effect in the cell nucleus. In all such studies, there was hardly any difference in effect between a total dose applied within a brief period, or over an extended period. If any, there was generally less of an effect for protracted doses. In other words, a nearly constant effect per rad was widely believed to apply over a span of dose rates, ranging from very low to very high rates.

It has long been known that in the cell nucleus, the DNA molecule which carries the genetic information is directly damaged by the impacting rays. As Petkau discovered, a very different, indirect damage-producing mechanism operates in the case of the cell membranes.

How Can Small Doses Be More Dangerous Than Large Ones?

In the cell fluid, which contains dissolved oxygen, the radiation can cause the formation of a highly-toxic, unstable form of oxygen. These so-called "free radicals" of O_2^- are attracted by the cell membrane, where they set off a chain reaction that successively oxidizes the molecules of the cell membrane; this weakens or even destroys the membrane. Thus, unlike the case of the cell nucleus, the damage is not the direct result of radiation; rather, it occurs indirectly, as the result of the "free radicals" created by the radiation.

- *Serious damage in the case of small, extended or chronic radiation doses:* The fewer "free radicals" present in the cell plasma, the greater their efficiency in producing damage. This is because the "free radicals" can deactivate each other to form ordinary oxygen molecules. Thus, the fewer "free radicals" created by the radiation in a given volume—and smaller doses create fewer of them—the greater their chances of reaching their goal, the cell wall, without first being subjected to a recombination.

- *Slighter damage with higher, short-lived radiation doses.* Conversely, the more "free radicals" are created by the radiation dose in a given volume of tissues—and a higher dose at one time will create more—the faster they recombine and become ineffective before they can reach and damage the membrane.
- In addition, however, there is a further effect. Cell membranes generate an electrical field in the cell plasma which attracts negatively-charged molecules such as the highly-toxic "free radical" $)^2$. Computer calculations have shown that the greater the concentration of "free radicals," the weaker the attraction by the electric field.[135] Thus, if the concentration of radicals is high, the "free radicals" are even less capable of reaching the cell wall, than if the immediate concentration of radicals is very low.

Thus, in contrast to the cell nucleus, the cell membrane is less heavily damaged per unit of absorbed radiation by a dense dose of ionizing radiation, such as alpha radiation or the short, intensive radiation burst of a medical X-ray machine, than by extended or chronic low-level natural radiation, fallout or nuclear power plant emissions.

Sternglass was the first to recognize this as early as 1974.[170] He deserves credit for having searched, compiled and analyzed radiobiological research work in literature. His results show the workings of the Petkau Effect in biological systems.[171, 172, 174] The BEIR III report of 1980 was for the first time forced to mention Sternglass—also the Petkau Effect was mentioned, for the first time.[132] Interestingly, the ICRP and the UNSCEAR have yet to do so.

Due to the Petkau Effect and according to Sternglass, small and minuscule extended radiation doses, such as those which affect living things following fallout or the emissions of nuclear power plants, would be 100 or 1000 times more dangerous than that experienced by the A-bomb victims in Japan or from thousands of experiments with animals.

Damage to Cell Nuclei Considered, Damage to Cell Walls Ignored

Originally, it was believed that radioactivity only caused genetic damage. Because the genetic material is located in the cell nucleus, there was an absolute fixation on that part of the cell. Moreover, experiments with animals indicated that in regard to genetic damage, the use of a linear extrapolation from higher to lower doses seemed to be on the safe side—i.e., seemed to avoid underestimating the genetic risk.[52] In a few experiments, a reduction of genetically-damaging radiation effects had even been determined at lower doses, i.e., an extended exposure produced less damage.[52, 148] This was explained as a possible regeneration in cases where the radiation was administered in low partial doses over an extended period, i.e., improved repair opportunities for the genetic material in the cell nucleus was assumed.[52]

Later, when it became necessary to recognize radiation-induced cancer as a major effect of radiation, the damage was viewed once again as likely to occur primarily in the cell nucleus itself, even though the exact mechanism of the genesis of cancer due to radiation was not, and is still not, known. Analogous to the risk of genetic damage, it was again assumed that a linear extrapolation from higher to lower doses would be on the safe side.[7, 52] There was much self-satisfied talk of conservative practice, i.e., overestimation of the risks of radiation. The linear relation could even be shown to hold true for the risk of cancer to babies X-rayed in their mothers' wombs.[208] As late as 1973, a study of all extant animal experiments on radiation-induced cancer showed that with varying dose rates, the same trend toward a lower risk of protracted dose rates applied as in the case of genetic damage.[118, 171] However, all these studies had been carried out with dose rates which were more than 1000 times greater than are found in natural radiation.

The correctness of the conservative, linear correlation seemed to be supported by all other research results with human population groups that had been exposed to medical radiation or nuclear explosions, as was reported in BEIR I in 1972.[17, 171] The relative cancer risk seemed to remain constant across a range of six orders of magnitude, from the low dose rates common in nuclear medicine (0.01 rads/min.) to the high ones generated directly by A-bomb explosions (10,000 rads/min.). This was true as well for the dose needed to double the risk of cancer, a range of 10-100 rads (0.10 – 1 Gy). A similar range had also been found for the doubling dose that causes genetic damage. This again supported the evidence for a linear cancer risk.[171]

All these findings appeared to indicate that low-level exposure to radiation at approximately the natural radiation level could hardly be dangerous—including, obviously, fallouts and the emissions of nuclear power plants. The only exception seemed to be unborn babies, for whom extensive studies revealed that following diagnostic X-rays, the doubling doses for leukemia, cancer and other causes of death was less by factors of between ten and 100, than was the case for adults.

One-sided Research: Mistaken Conclusions

The following table shows that until now research has primarily been performed at the wrong, high doses and dose rates: the blast of the A-bomb, medical applications, animal experiments. The dose rates ranged between one and 1000 rads/minute. The conclusions drawn for the case of very low doses and dose rates were erroneous. They did not apply to environmental risks due to natural radiation, fallout from nuclear weapons tests and the emissions of nuclear facilities, as well as on-the-job risks of handling ionizing radiation. In all these cases, the range of the dose rates is many orders of magnitude less—between 0.000,000,1 and 0.0001 rads/ minute.[170, 173]

The table below gives an overview of dose-rate ranges:

Dose Rates, in rads/minute			
10^{6}	1,000,000		Electron-beam therapy machines
10^{4}	10,000		[Nuclear blast,
10^{3}	1,000		[direct radiation
10^{2}	100		
1	1		Medical (diagnostic)
10^{-2}		0.01	Medical (nuclear isotopes)
10^{-4}		0.0001	
10^{-6}		0.000,001	Fallout and natural
10^{-8}		0.000,000,01	radiation

Petkau Effect Confirmed in Living Systems as Well?

Numerous scientific studies of the past twelve years have shown that indirect cell membrane damage must also be effective in biological systems, even at the most minimal doses of 10-100 mrads (0.1-1 mGy), i.e., in the range of natural radiation, fallout and an average dose of emissions from a nuclear power plant. The results of earlier studies are only now emerging as explicable and credible, in the light of the Petkau Effect. Sternglass has cited many such studies.[170, 171, 172, 174] Let us review just a few of them here:

- It has been shown by means of microorganisms that the Petkau Effect can be ascertained in living cells (W. S. Chelack).[43] By replacing the oxygen dissolved in the cells with nitrogen, the dose required to destroy the cell membrane would be increased considerably. This confirms the decisive role of oxygen—or rather of the free O_2^-.
- BEIR III 1980.[32, 51] Here, a series of studies is referenced that show the possible means to protect the membrane by various enzymes and substances which deactivate the free radical (O_2^-). These results provide indirect proof for the potential of free-radical damage to the membranes in living systems.
- It was discovered in rats that the lower the strontium 90 concentration (and hence the dose rate) in the bone marrow, the higher the damage per millirad to the bone (W. T. Stokke).[181] The concentrations of strontium 90 per gram of body-weight in this experiment were in the same range, as in the bodies of those new-born babies at the height of the A-bomb testing period.
- Studies were conducted on human blood cells taken from occupationally-exposed personnel (radiologists, X-ray technicians, etc.) at the University of California (E. G. Scott).[151] The membranes of the blood cells were much more permeable, or severely damaged, for exposed than for unexposed persons. Again, as found by Petkau and Stokke, the greatest increase in damage per unit dose was found with the smallest total doses. The percentage change per rad at low doses was shown to be greater by a factor of 100, than might have been expected according to studies of higher doses.
- B. Shapiro and G. Kollmann.[153] Petkau's discovery had actually been anticipated as early as 1968, although from the other extreme of very high doses and dose rates. Shapiro found that at a high dose rate of 1900 rads/minute (19 Gy per minute), comparable to exposure from an A-bomb explosion, the total dose of a high 2000 rads (20 Gy) was needed to damage the membranes of blood cells.
- In rats that had inhaled plutonium dust during a long-term experiment, the doubling dose for cancer was only 180 mrads (1.8 mGy) (C.L. Sanders.[149] If, however, from the result of the highest dose of

395 rads, one were to calculate the doubling dose by extrapolating linearly downward toward the zero point, as is still commonly done today (and is considered conservative), one would obtain a doubling dose of 34 rads. One would thus have erred by a factor of 190, and hence, have underestimated the risk of radiation in the low-dose range by that same factor. The knowledgeable reader may care to take a look at the graph below, which shows the Sanders data.

- *In hamsters,* it was also possible to observe a greater effect per rad at smaller doses (J.B. Little).[108] Polonium 210 was administered to the lungs of the animals; the greatest increases in cancer rates per unit dose occurred with the smallest radiation doses, in accordance with the observations made by Petkau, Chelack, Stokke, Scott and Sanders.

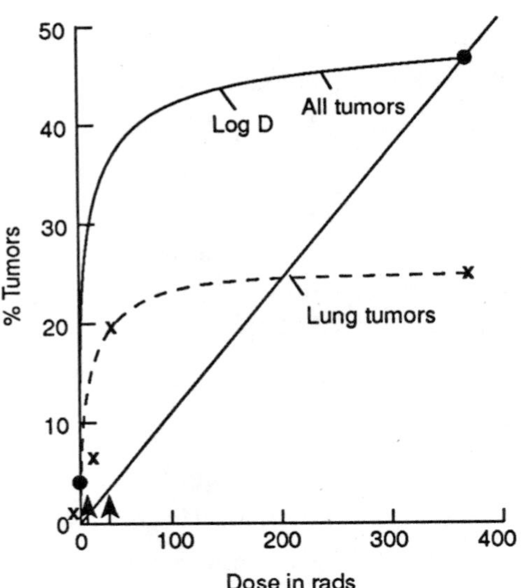

The solid line shows the relationship of *all* cancer cases to the dose applied. The broken-line curve applies only to lung cancer. Note the very steep climb for small doses, as is to be expected for the indirect effect of cell-membrane damage. The diagonal broken line shows the linear extrapolation from the high dose of 395 rads.

It was possible to cause 15 times as much cancer in the lung if polonium was allowed to have its effect distributed over the course of 15 weeks than if the entire dose was administered at once (R.B. McCandle). [110]

- *Beir III 1980* notes a very important observation by Petkau.[32] By contrast to the protective effect of enzymes found if the irradiation of an artificial cell membrane occurred externally by means of gamma radiation, no protective effect could be observed in the case of internal irradiation by means of tritium. The BEIR subcommittee considered it an urgent matter to ascertain whether this reaction might also be observed in living systems.

Confirmation of the Petkau Effect in Humans?

There are an additional number of indications that the linear dose-effect curve does not correctly reflect the risks of low-level radiation for humans. The risks are in fact underestimated many times. The following studies should be mentioned in this regard:

- Personnel at the Hanford plutonium facility in Washington State indicated such a high rate of cancer in 1977, in spite of only minimal, low-level radiation exposures, that researchers Mancuso, Stewart and Kneale recommended a reduction of the maximum dose for radiation workers by a factor of 20.[113] No holds were barred in the fight to refute the study. Even the values ascertained were disputed. As always, the strictness of the supervision of such workers was stressed.
- Najaran and Colton in 1978, obtained similar results for U.S.A. Navy Shipyard workers who repaired nuclear submarines in Portsmouth, New Hampshire.[23, 156] A leukemia rate 5.6 times higher than for that of non-exposed workers was found.
- J. T. Gentry, in 1956, found that increases in mortality of newborn infants in New York State were due to various developmental deficiencies of between 20% and 40% in areas with naturally highly-radioactive rock (due to uranium and thorium content).[55] In agreement with Stokke, this yields an increase in the effect of

1% per mrad (0.001 rads). This result also matches Scott's work on damage to the blood cells of occupationally-exposed personnel.[174]

- According to J. P. Wesley, there is a close correlation between stillbirths with visible deformities, and the intensity of global cosmic radiation.[156, 210] According to his data, published as early as 1960, there are 1.8 deformities per 1000 births at the equator, while the rate is 5 per 1000 in areas above 50° north parallel. This finding became acceptable/comprehensible only after the discovery of the Petkau Effect. The great increase in stillbirths could not have been explained by an effect on the DNA in the cell nucleus (genetic damage).[174]
- Barcinsky and Costa Ribeiro have found numerous chromosome variations in the blood of persons living on thorium-rich monocyte sand soil; the investigation involved both local residents and monocyte sand workers.[44, 156] A ten-fold increase in the radioactive lead 212 content in the air causes an increase in chromosome defects of 0.9%, to 2%. An additional ten-fold increase of the lead 212 concentration, by contrast, multiplies the chromosome defects by only another 0.57%. Again, a great effect per unit dose was observed at low doses.[174]
- In 1984, *The New Scientist* reported on internal studies by the United States Department of Energy (DOE) on workers in twelve different nuclear facilities.[132] Nine of the studies found up to 50% higher leukemia rates, higher than average rates for lung, lymph and brain cancer, as well as malignant tumors in the digestive organs. Common diseases of the respiratory system were likewise more frequent. A study of 2,529 workers in various DOE facilities receiving more than 5 rem (50 mSv) of radiation per year, revealed that the cancer rate was three times as high as expected.

The following was stated:

This project provides an opportunity to evaluate scientifically the risks to human health, particularly the cancer risks, of protracted exposure to low doses of ionizing radiation. Radiobiological and epidemiological evidence suggests these risks may differ from those associated with simple or fractioned high doses on which current risk estimates are based.

According to studies by Petkau, Stokke, Scott, Sanders and Little, a total dose of 0.1-0.2 rads (0.001 - 0.002 Gy) of natural radiation, fallout and nuclear facilities emissions is sufficient to produce clearly detectable damage to the cell membrane.[177]

The dose rates of medical X-ray technology, by contrast, require the high total dose of 100-200 rads (1-2 Gy) to double the damage. Most existing studies on radiation damage to humans and animals are based on these high dose rates, or even higher ones. It seems therefore that the harmfulness of natural radiation has been underestimated by a factor of 100 or 1000. Rather than the hoped-for level of tolerance, as indicated by curves c and d in the graph below, or even the conservative linear correlation (curve b), the new research results show a "supralinear" curve a—i.e., one that arches upward.[174] This means very steeply climbing cell-membrane damage in the low-dose range, and a flattening of the curve later on.

General Consequences of the Petkau Effect

Radiation damages the large surface of the cell—not only the small cell nucleus, as was previously thought. This indirect damage to cell mem-

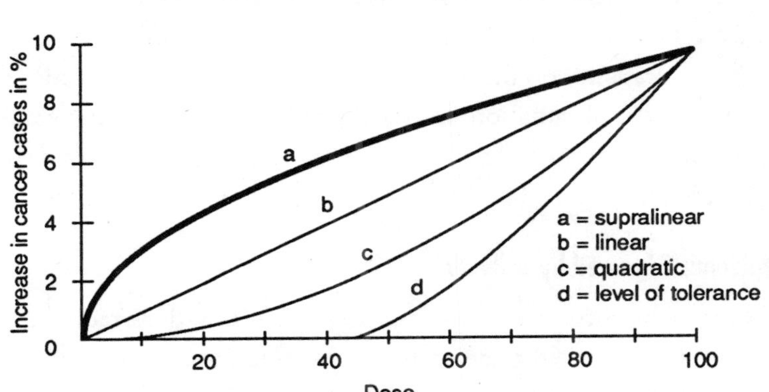

branes manifests itself more strongly at lower doses, leading to a concave downward dose-effect curve (a) which climbs more rapidly in the lower dose range than would a linear curve (b).

The aforementioned Hedi Fritz-Niggli, Director of the Radiobiological Institute of the University of Zurich attended (as did the author) a symposium at the University of Bern in June, 1986. In discussion, she admitted that such a concave downward curve was possible.

Cell Membrane Damage and Health

The Petkau Effect seems to cause damage to those cells which are responsible for the body's resistance to disease. This increases the risk of infection. Viruses, bacteria and cancer cells have an easier time reproducing. Developing, unborn life, such as babies in the womb whose immune systems are not yet fully developed, are evidently especially threatened.

Thus, small radiation doses can ultimately lead to damage which had not previously been suspected. This includes infectious diseases, such as influenza and pneumonia, as well as all such diseases of aging, such as emphysema, cardiac disease, thyroid disease and diabetes. Particularly serious, is brain damage to the developing embryo and fetus, which obviously leads to reduced mental abilities.

Prior to the discovery of the Petkau Effect, it was impossible to explain the numerous and multifarious statistics which indicated increased risks due to fallout and the emissions of nuclear plants. After all, the radiation doses due to these sources only amounted to 10-100 mrads per year (0.01-0.1 rads), as calculated for the fission products cesium 137, strontium 90 and 89, and iodine 131, ingested in food, milk and drinking water.

Also, statistics on the effects of natural radiation and occupational exposure to artificial radiation, revealed risk levels much higher than had been previously assumed. As a result of the Petkau Effect, these calculations became plausible.

AIDS Epidemic Triggered by Fallout?

The discovery by Stokke and his co-workers that low-dose radiation from radioactive strontium 90 appears to produce damage to the cells of the immune system has lead to the hypothesis that the AIDS epidemic may have been triggered by the massive fallout from atmospheric A-bomb tests in the 1950s and early 1960s. This possibility, suggested by Sternglass and Scheer, may explain two puzzling aspects of the epidemics origin: the precipitous increase of AIDS cases in 1980-1982, and its initially high concentration in central Africa, the Caribbean and the East and West coasts of the U.S.A.[222,223]

These facts may be explained as follows:

1. During atmospheric A-bomb tests, the beta radiation caused by strontium 90, and other bone-seeking radioisotopes, caused the mutation of an existing HIV-like human or animal virus of low virulence to mutate into a deadly form. This may be the origin of the HI-virus. It may however have a much older origin. In 1987, the French AIDS researcher Professor Montagnier, claimed the virus first appeared in several monkey species and then spread among central African peoples.
2. During the period of atmospheric A-bomb tests, which reached its height in 1963, a large group of humans were born with weakened immune systems due to the irradiation of the immune system of the bone marrow by strontium 90 and 89.

Some 18 years later, i.e., around 1980-1982, a strong increase of AIDS cases ensued concerning those children who carried the mutated virus, and who were now sexually mature. With sexual maturity, the impaired immune systems became exposed to sexually transmissible diseases. This caused the multiplication of the T-cells, which are vital to the immune defense, and hence the activation of the latent HI-virus. The HI-virus uses precisely these T-cells to multiply, as a result of which the cells die. Under favorable conditions, the virus is able to spread: frequent sexual contact, or directly through the blood system.[227] Some 90% of fallout, including strontium 90, comes to earth through precipitation. *According to the new hypothesis, the AIDS epidemic began in areas of high rainfall* such as central Africa, the Caribbean, and around the geographic latitude of the Pacific A-bomb test sites. It increased most strongly in the high-rainfall areas of the East and West coasts of the United States. There were far fewer AIDS cases in drier regions, such as Northern and southern Africa, and in the plains of North America's interior.

Sternglass and Scheer also support their hypothesis on the basis of the strontium 90 contents of food. Thus, Southeast Asia had few AIDS cases—although it, too, is a high-rainfall area—since rice and fish have much lower strontium-to-calcium ratios than milk, bread, fruit, potatoes and root vegetables, which predominate in the U.S.A., Caribbean and African diets. (The higher the strontium-calcium ratio is in the diet, the more strontium will be incorporated into the bones.) According to the 1962 UNSCEAR report, the young people with the highest strontium 90 concentrations among 22 countries studied in 1957, were to be found in tropical central Africa.

Unfortunately, it is not possible to list all the arguments that the authors cite in support of their AIDS hypothesis. Nonetheless, two important studies should be mentioned:

- Recent laboratory studies have shown that bone-seeking radioisotopes, such as strontium 90 and 89, injected into mice will disable the proper function of certain white blood cells formed in the bone marrow. These are the so-called "natural killer" (NK) cells, which serve as a defense against disease-causing cells and viruses.[224] We are coming to understand the manner in which these killer-cells destroy their targets in the service of our immune defense systems.
- In addition—and this has also been documented in laboratory studies—very small doses of strontium 90 in the range of 10-100 mrem cause a significant reduction of bone-marrow cells in rodents.[181] This effect can be the result of indirect damage caused by free O_2 radicals within the cell membranes. With fewer bone marrow cells, the bone marrow will of course produce fewer of the cells vital to the immune defense system.

Sternglass and Scheer assume that one of the most serious consequences of fallout is the unexpected long-term effect of immune deficiency acquired in the mother's womb. It may remain unnoticed until years later when the body is attacked by infectious disease. If, in addition, radioactive effects caused by isotopes such as strontium 90 in the bone marrow cause an existing virus to mutate—or if a virus was already present in the body—the entire immune system may collapse, causing death due to infectious disease or cancer. This is what happened with AIDS.

Sternglass was the first to point out that existing radiation-protection law is based on the false assumption that radiation-induced cancer occurs only as the result of a malignant mutation in the cell nucleus. Yet, now it has been shown, particularly with AIDS patients, how important an intact immune system is for defense against cancer. When it collapses, not only infectious disease but also cancers, particularly lymphatic cancer and Kaposi's sarcoma, a serious skin cancer, develop.

Since the discovery of the Petkau Effect in 1972, Sternglass hypothesized (again, he was the first to do so) that low-level radiation would magnify all health risks. New research has been ever reenforcing the efficiency of the Petkau Effect. Thus Sternglass refers to an article by Marx which summarizes the important role of the free oxygen radical O_2^- in the genesis of many diseases.[226] Of course, such radicals do not occur solely as the result of radiation, but also as the result of biochemical reactions. For the first time these concrete correlations with the free oxygen radical are being recognized.

Sternglass notes, interestingly, that the famous dissident Russian Nobel laureate and Professor, Andrei Sakharov, predicted, as early as 1958, the decisive role of the free radical in causing health damage and the overall weakening of the human immune system.[223, 226] He also feared that the fallout would cause viruses to mutate increasingly into virulent or aggressive forms, including worldwide epidemics of new infectious diseases.

The Radiation-Protection Agencies' View of the Petkau Effect

Both the ICRP and the UNSCEAR have continued to ignore this fundamental discovery. However, the BEIR Subcommittee (BEIR III 1980), at the urging of Sternglass, finally agreed to contact Petkau directly. Thus, the importance of his discovery and, the significance of the free-radical effect was confirmed. It reported:

> The experimentally-documented effects of ionizing radiation on cell membranes provide an alternative or conjunctive damage mechanism, in addition to the effects on DNA which are generally accepted as the primary modes of damage in biological systems . . . Although it is well recognized that membrane integrity is essential for normal cell function, there is inadequate basic understanding of membrane structure and function on which to base a

> detailed theory of radiation-induced damage mechanism. The role of radiation damage of membranes in the induction of pathological effects in living systems has not been established, although possible connections to carcinogenesis, autoimmune diseases and aging have been proposed.[32]

These authorities now concede that a free-radical damage mechanism must also be assumed in biological systems "at lower dose rates approaching natural background."[32] Numerous studies with supporting data are referenced. One by T. E. Fritz demonstrates the increased efficiency of small dose rates of gamma radiation in the induction of leukemia in dogs.[32] The studies of Mancuso, Stewart and Kneale are cited without comment. They discovered increased cancer rates among workers at the Hanford, Washington, plutonium plant. A greater effect per unit dose at low doses than at higher doses was revealed.

Following its discussion of the free-radical effect, the BEIR III report of 1980 states:

> An inverse relationship between dose rate and the induction of damage to model membranes systems and the possible relationship of such alterations in biomembranes to carcinogenesis suggests that this phenomenon may be involved in low-dose or low-dose-rate effects in living systems. Thus, there is a need for further studies in this field.[32]

Petkau also assumes that there is a connection between cancer and cell membrane damage in living cells.[219, 220]

In 1989, when the laboratory management cut off all funds for his research team, Petkau resigned his post as head of the Medical Biophysics Branch of the Canadian Atomic Energy research laboratory in Pinawa, Manitoba. Presently an associate professor in the Department of Radiology of the University of Manitoba and the author of 92 papers in the field of radiational biology, he has opened his own medical practice.

Conclusions Drawn from the Petkau Effect

Research to date supports the assumption that indirect chemical damage to the cell membrane *in the low-level dose range* is much more significant, than the direct effect of radiation on the cell nucleus and its genetic material (DNA). There seems to be more effective repair mechanisms in the cell nucleus than in the cell membranes. This is in accordance with the natural goal of evolution—development of life to ever higher levels. Most important, from the point of view of evolution, is to protect the genes in the cell nucleus by means of highly-developed repair mechanisms. Only in this way will it be possible for nature to protect the gene pool against the highly damaging background radiation, and thus maintain the relatively great stability of species over millions of years.[172]

By contrast to the genetic material, individual members of populations (plant and animal species, as well as humans) are not that important for evolution. On the contrary, their continual death and replacement by reproduction is a necessary evolutionary process. Repair mechanisms in cell membranes do not need to be, and in fact can not be, as effective. Damage to cell membranes only cause diseases, but not genetic damage. Once a member of a population has fulfilled its reproductive task, it is no longer needed, so that extending life beyond reproductive age is not absolutely necessary. Yet, rapid adaptation to toxic environmental substances is impossible for humans, due to the long period required for generational change.

If, therefore, humans wish to enjoy a long life, and health as free as possible from congenital defects, cancer, heart disease and other chronic conditions, the natural radioactive exposure of the environment should under no circumstances be increased. Likewise, we cannot afford to add carcinogenic and genetically damaging chemicals to our air, water and food.[172]

The Bombshell of 1981: The Most Important Radiation-Protection Data for Humanity Are Wrong

"The analysis of the survivors of the atomic bombings of Hiroshima and Nagasaki exceeds all other studies by far, both in terms of quality and of quantity," experts still said in 1981.[48] Only a few other population groups, such as groups exposed to medical X-rays, were large enough to receive documentation. The study of the A-bomb survivors had surpassed all others in the reliability of the data, and became a kind of Bible for the setting of radiation standards.

This "Bible" is based on the famous "Tentative Dose Estimate,

1965," or TD65, where the radiation doses received by the Japanese A-bomb victims, calculated after the fact, are set down. In order to verify this data, a special test explosion was even conducted in Nevada.[116]

Scientists of the U.S.A. weapons laboratories recently recalculated the radiation fields of the two A-bombs—"Little Man," dropped on Hiroshima, and "Fat Man," detonated over Nagasaki. They found that the neutron radiation had been overestimated by a factor of 6 to 10, whereas the gamma radiation had been only slightly underestimated. As a result, the basis of the existing radiation-protection standards has been shaken—the currently-estimated cancer risk has been proved to be too low, even for short bursts of high-dose radiation.

Opinion as to this underestimation has diverged.[116, 117, 139, 202] In any case, the cancer risk estimates in the BEIR III Report of 1980 are now invalid.[202] Recalculation of the data resulted in unexpected difficulties. When the U.S.A. National Commission on Radiation-Protection requested detailed documentation of the TD65 study, it was discovered that some of the files had found their way into the waste basket—in other words, the most important data in existence for human radiation-protection has disappeared![116] It is not yet known how long it will take for the process of recalculating the doses to be completed. When the results are in they are unlikely to be accurate. Already, there are those who say that the mistakes that have been made will have no effect on the result![95] The 1982 UNSCEAR report estimates that the risk of cancer as derived from the Japanese figures is actually twice as great.[202] For this reason alone, all dose limits should be reduced by half—but then nuclear power plants could not longer operate economically.

Wrong Again by a Factor of Ten?

A truly serious fact has begun to be stressed again recently: it was only in October 1950, 5 years after the detonations in Hiroshima and Nagasaki, that the study of the Japanese survivors began. According to the respected scientists Stewart and Kneale, two important factors were therefore, not even taken into account:

- During the first five years (i.e., before the study began), those whose health was weakest died in large numbers. Many of these survivors, however, died of delayed effects of the bombs, although this could no longer be ascertained.[3] Such effects following other catastrophes are well known. In Hiroshima and Nagasaki, as in other disaster areas, water, food, medicine and housing were initially in short supply, and poor sanitary conditions prevailed. The population was thus subjected to a rigorous selection process, so that, as UNSCEAR pointed out in 1964, only the strong and healthy survived.[197]

- The second factor cited by Stewart and Kneale is the possible long-term effect of radiation after 1950. Many survivors suffered radiation damage to the bone marrow, which plays an important part in the body's immune system. This damage also causes a disease of the blood known as aplastic anemia. This in turn causes increased susceptibility to infectious diseases of all types, including tuberculosis, pneumonia, bronchitis and infections of the kidney. Those who died of such diseases are not, however, reflected in the cancer and leukemia statistics, although they, too are victims of radiation. Stewart and Kneale found their hypothesis confirmed on the basis of statistics on the non-cancer-related deaths of the Japanese.[3] Still, the Radiation Effects Research Foundation (RERF), like the ICRP, merely lists all unusual diseases of the blood as mistaken diagnoses. It should also be pointed out that the RERF denied Stewart and Kneale access to important data bearing on this problem.

Taking these two neglected factors into account, Stewart and Kneale consider that probably ten times as many Japanese died of the effects of radiation than has previously been assumed; two thirds of these were victims of diseases other than cancer.[3]

Demand for Independent Agencies and Independent Studies in Radiation-Protection

Robert Alvarez, director of the Washington, DC-based Environmental Policy Institute's Nuclear Power and Weapons Project, demanded in the *Bulletin of Atomic Scientists* of October 1984 that since the Japanese A-bomb study could no longer evidently provide the necessary data for the setting of low-level radiation standards, a large-scale study should be undertaken out of the 600,000 occupationally-exposed individuals employed in government nuclear facilities since the 1940s. These employees had been subjected to low-level radiation only, moreover, radiation carefully measured and recorded for each individual.

At the same time, Alvarez demanded that such studies be taken out of the hands of the DOE and assigned to a public health agency, such as the National Institute of Occupational Safety and health. After all, the DOE's job is to develop nuclear weapons and promote nuclear power.

Finally, Alvarez demanded that the Japanese A-bomb study, including the official estimates of radiation risks, be reviewed by independent scientists.[3]

Faith in current radiation-protection standards is thus shared by an ever decreasing number of influential scientists. In direct violation of the protection of human life, these standards are being manipulated to suit economic concerns. Not only the public, but also the many workers in the nuclear industry are apparently expendable.

The ICRP Loses its Last Shred of Credibility

A New Structure for Radiation-Protection Legislation

Instead of reducing existing dose limits by factors of at least two - 20, as is being demanded in professional circles, the ICRP continues to stubbornly hold fast to the established standards. As early as 1972, the National Academy of Sciences has recommended that the limit for the population at large be reduced from 170 mrem/year to only a few millirems. Their argument was that every individual should have the right of protection from cancer induced by artificial radioactivity.

Instead, the ICRP unscrupulously interpreted the introduction of a new framework for radiation-protection legislation—the so-called weighting factor—to promote an increased maximum permissible dose in individual body organs.

Since the inception of nuclear power, radiation-protection legislation has been based on disastrously false premises, as I had warned in 1972, in *Die sanften Morder—Atomkraftwerke demaskiert [The Gentle Killers—Nuclear Power Plants Unmasked].*[61] At that time, the ICRP described the situation as follows:

> It would be generally agreed that when the whole of the body is uniformly irradiated so that the tissue dose is sensibly uniform the total cancer risk will be some sort of sum of the individual risk for each organ separately. Yet at present the dose limit for exposure of the whole body is determined by the tissue dose in the bone marrow and gonads without reference to the cancer risk from the concomitant exposure of the rest of the body.[61, 74]

Thus, the ICRP actually had admitted that meaningful risk calculations for cancer are not even possible. The responsible authorities kept this information from the public.

In 1969, the ICRP had described a more accurate approach, in which total cancer risk no longer depended on the dose to a single critical tissue, but rather on the sum of individual risks of 27 organs.[77] Since, however, the documentation required for this purpose was not yet available (even today it is far from complete), the existing flawed framework was maintained.[70] Since this framework underestimates the risk of cancer and is, moreover, easy to apply in practice, it has served to smooth the way for nuclear power.

Since, however, a growing body of research continued to show that the risk of cancer had been underestimated, the ICRP in 1977 proposed a new approach to risk calculations involving certain weighting factors for individual organs. Yet, the exact relative radiation-sensitivity of most organs, and hence the risk of cancer developed in a given organ, is still not sufficiently well-known.

In Bulletin No. 124, the Swiss Association for Atomic Energy (SVA) gives the following overview of the weighting factors for individual organs, in percent, (see figure on following page).[54]

All this looks very benign and conscientious to the lay person. The sum of weighting factors for genetic damage and cancer (including leukemia) is 100%, of which 25% is related to genetic damage and 75% to cancer. This 75% again breaks down according to various organs. The maximum whole-body doses of five rem/year for occupational exposure and 500 mrem/year maximum permissable dose for general-population exposure are maintained.

Weighting Factors (ICRP 26)

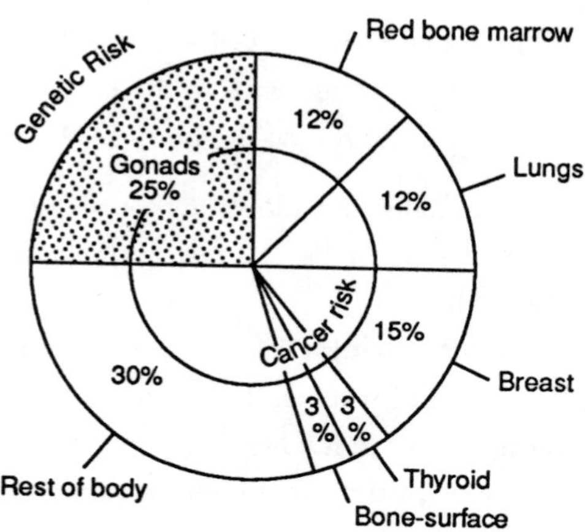

The Monstrous Trick

In 1978, shortly after the ICRP proposal was published, Karl Z. Morgan, former chairman of the ICRP and first president of the Health Physics Society, also known as "the father of radiation-protection", criticized this innovation.[120] Morgan pointed out that it was being used to sharply increase the maximum permissible dose limits for individual body organs. Naturally the maximum permissible concentration of radionuclides in the air, water and food would also sharply increase, except in rare cases when their distribution is relatively uniform throughout the body.[120]

Morgan again vigorously criticized the ICRP, in 1986, complaining that the Commission was permitting increased radiation exposure to crucial organs, despite growing evidence that the risks had been seriously underestimated.[215]

The following table shows the ICRP recommendations for occupationally-exposed individuals:

Organ	MPD/yr.[120, 71] old	MPD/yr. [101, 86] new	Weighting Factor in %[86]
Full body	5 rem	5 rem	100
Gonads:	5	20	25%

Organ	MPD/yr.[120, 71] old	MPD/yr. [101, 86] new	Weighting Factor in %[86]
Red bone marrow:	5	42	12%
Lungs:	15	42	12%
Thyroid	30	50	3%
Bones:	30	50	3%
Breast:	15	32	15%
Skin:	30		
Rest:	15	17	30%

*MPD/yr.: Maximum permissible dose per year.

These increased permissible organ limits are very serious, since they allow large discharges of the most dangerous radioactive fission products, such as iodine 131 and strontium 90 to contaminate the environment. Although some nations have accepted the new ICRP recommendations, others have not. The ultimate outcome of this effort to permit increased exposure of workers and the general public is still uncertain.

Clearly, attempts are being made to generally increase the organ doses for radiation-exposed employees to 50 rems (with the exceptions of skin and eyes). Given the present rule, allowable doses to individuals in the general population are one-tenth of occupational doses, the new ICRP proposal can raise the maximum permissible doses for key organs to 5 rems for the general public. This is obviously an attempt to pave the way for the ever-increasing radioactive contamination of the environment—more nuclear power plants, more accidents, more catastrophes, and more nuclear fuel reprocessing.

One of the main problems with the neat-looking pie of weighting factors is that cancer mortality and genetic damage are evaluated together, whereas no ethically responsible common scale of damage is possible. Genetic damage can cause the most varied kinds of suffering—life-long handicaps, mental deficiencies, medical problems. These consequences are obviously so different that it is invalid to create a common denominator. In fact the ICRP itself, recognized this problem in 1969, stating that insufficient experience in regard to genetic damage is available, in addition to the lack of specific scientific analysis of the relative harmfulness of genetic damage and cancer.[78] The ICRP has now unscrupulously discarded these misgivings.

Natural Radiation Suddenly Simply Doubled

The international radiation-protection agencies are well-known for their repeated mistakes in risk calculation. The quantitative limits stated in rem, upon which risk calculations are based, are, after all, nothing but well-meaning guesses at best. These limits are clearly not licenses to release ever-increasing amounts of artificial radioactivity into the environment. However, utilizing atomic energy for civilian and military purposes means exactly that.

In 1982, in an effort to minimize the significance of the increase in permissible artificial dose, the UNSCEAR simply *doubled* the estimate of normal natural radiation exposure. They explained:

> Until recently, the annual whole-body dose from natural sources was estimated at 1 mSv (100 mrems). However, in the 1982 UNSCEAR report, the annual equivalency dose is estimated at 2 mSv (200 mrems). This higher estimate is the result of the addition of the equivalency dose which is caused by the exposure of the lungs to the fission products of radon and thoron, principally in the air inside houses, rather than of steady irradiation of the entire body by the components of natural radiation.[89]

The doubling of the natural radiation dose is thus based primarily on the replacement of the existing faulty radiation-protection framework. Its false "whole-body dose" and new principle of the "whole-body equivalence dose" uses the "weighted factors" proposed by the ICRP to permit larger organ doses. Another factor is the assumption that radon may affect us more strongly than had been assumed. As a result, exposures to alpha, beta and gamma radiation in the individual organs such as the lungs must be taken into account, even though the gonads and the bone-marrow are not affected significantly by radon. By utilizing the proposed new radiation-protection methodology, the effect of radon is taken into account, conveniently lessening the significance of artificial-radiation releases from nuclear facilities.[203]

Protection from Natural Radiation Demanded

It is therefore not surprising that in 1984, the ICRP made a sudden 180°turn and began to discuss protection from natural radiation.[90] In recent years, they said, it had been realized to an increasing degree that human activity could give rise to higher natural radiation exposure. In particular, radon, a radioactive inert gas formed in the soil by the radioactive decomposition of uranium 238 and thorium 232, seeps into houses where it may collect due to poor ventilation. Very different quantities of radon can be found in different homes, depending on the composition of the soil and the radioactivity of the building material. A radon concentration increased by a factor of 1000 is possible in this manner.[90]

In cases of increases by factors of 100-1000, the only thing to do is move elsewhere, according to the ICRP.[90] The agency also recommends avoiding overly-radioactive building materials in the future.[91] The Commission likewise writes that future radiation-exposure situations would generally be handled by avoiding them—i.e., avoiding settlement of high-radiation regions which are not yet inhabited.[92]

The Swiss Radiation Monitoring Commission (KUR) estimated that the mean radiation exposure due to radon in single and multi-family dwellings in Switzerland was 125 mrem/year. This will cause 10-20 deaths per year per million inhabitants due to lung cancer.[102]

Bankruptcy Declaration

Many artificial radionuclides released during nuclear fission are highly dangerous alpha* and beta emitters. Depending on rather complex factors, they can lead to completely novel concentration mechanisms and unanticipated subtle effects upon a wide variety of organs and organic systems. It is absolutely impossible to ascertain by means of model-based calculations, the correct approach toward environmental contamination which would ensure that no one is subjected to impermissible exposure—aside from a ban on "permissible" toxins. Even if everything were known, monitoring would be impossible. In line with this fact, the ICRP explicitly recommends that the general population—as opposed to occupationally-exposed personnel—be monitored for the effectiveness of a limited equivalency dose, but only theoretically. The statement reads:

*Plutonium is an alpha-emitter.

Its effectiveness is being investigated by means of sampling procedures and statistical calculations, as well as by means of source control of expected radiation emitters; only in rare instances are samples taken of radiation exposure of individuals.[85]

There we see it. The population is not to be monitored individually. Radiation monitoring is generally to be accomplished in overview terms. If the general population—like the workers in power plants—had to run around with dose meters, take urine tests, and climb inside whole-body counters, nuclear power would be facing immediate shutdown.

There Are No Risk-Free Nuclear Waste Dumps

"Think extreme caution, right from the start!" is a slogan which has been used to make nuclear power palatable to the public. Certainly, efforts have been made to eliminate sources of danger, sometimes with enormous effort and expense. Unfortunately, however, this has been insufficient in preventing radioactive emissions and nuclear waste. Even after 30 years of nuclear waste production, experiments are still being conducted in order to learn how to store these wastes permanently.[64]

The dumping of low and medium-level radioactive wastes into oceans, once praised as safe, has largely been stopped. No one yet knows how far and for how long the release of radioactivity into the seawater will spread. In any case, it is accumulating in the plants and animals of the sea, and will eventually end up on our dinner plates.

The problem of highly-radioactive waste must be seen as fundamentally insoluble, since storage at calculable risk—lately, considered a satisfactory principle—is irresponsible. Absolute certainty as to the stability of the earth's crust cannot be assured, neither in regard to changes in the composition of the atomic wastes and their packaging, nor in regard to the drainage process and the ensuing migration of radioactivity to the surface of the earth and into the drinking water.

The uncertainties are confirmed, too, by the deepest holes ever drilled on earth, at the Kola Peninsula in northwestern Russia. In 1984, 12,000 meters—almost 7 1/2 miles—were reached, and even there, water and gas were found. We can no longer hope for compact virgin rock. The existence of cracks in rock at 3000 times the atmospheric pressure is something utterly unexpected.[38, 39] Attempts are therefore being made to date the water found at these depths, i.e., to find out how long it stays in the earth, and thus to determine the flow velocity.[122] Here, too, a favorable result offers no guarantees.

For this reason, there are now attempts to melt the wastes together with other substances so as to make them difficult to dissolve (for example, to form them into glass-like materials), and to then sheath them in additional material (for instance, steel or expandable clay).[97] It has even been suggested that the wastes be coated with gold in order to prevent their solution in water. In 1972, the Chairman of the United State's AEC, James Schlesinger, predicted that within ten years, nuclear wastes would be fired into the sun by rockets, rather than buried in underground caverns, as is presently being considered. It is no wonder that all but three states in the U.S.A. have rejected the storage of nuclear wastes within their borders, according to the government study, Global 2000.[56]

A few countries have scaled down their demands. Sweden, for instance, has amended its laws so that it no longer requires "safe" storage, but only storage at an acceptable risk. In Switzerland, too, standards have been lowered. The National Cooperative for the Storage of Radioactive Waste (NAGRA, in the German abbreviation) was given an assignment in 1978, to submit by 1985 a plan entitled *Gewähr* (*Guarantee*); it was to provide for permanent, safe disposal and final storage. By 1981, this task was reduced to outlining a solution, via a model project. Finally, in October 1982, it was decided that *Gewähr* should eliminate all doubts as to the implementation of waste disposal, and should suggest a model site.[212] There is no more talk of guarantees, only of "realistically-assumable circumstances under which the protection of humans and the environment will be provided."[121] As NAGRA adds:

> Although radioactive substances can in fact move out of permanent disposal sites into the general environment, the radiation exposure will under all circumstances be far below the values established by the authorities.[121]

According to Swiss guidelines, these limits are ten mrems. Thus, there is no real confidence in implementing leak-proof final disposal. Mar-cel Burri, professor of geology at the University of Lausanne, pointed out in 1981, that geologists are not even capable of predicting what's going to happen over the course of 25 years at a depth of a few hundred meters; he cited as an example, the settling of the storage wall at the Tseuzier power plant in Switzerland.[42] Burri said that the same is true regarding the nuclear waste problem: "Once again, geological reports are supposed to provide reliable information regarding geological conditions at several thousand meters' in depth over the next few hundred thousand years. Shouldn't geologists admit that they can't fill the bill¿"[42]

Experts were also wrong about the so-called "safe" final disposal site at Gorleben, Germany. In 1984, the fundamental suitability of rock salt as a natural barrier for a final disposal site was questioned. Scientists in the U.S.A. demonstrated that rock salt is most effectively broken down by radiolysis at precisely those temperatures (radioactive waste heats its surroundings) which are indigenous to a rock salt final disposal pit during the first 50 to 100 years of operation—150-175° C.[209] This destruction of the rock salt (NaCl) by radiation leads to the formation of colloidal sodium (Na) and gaseous chlorine (Cl). Water of crystallization is liberated, so that the waste will ultimately be surrounded by a dangerous mixture of gaseous chlorine and hydrogen, as well as sodium and lye. A mockery of the pursuit of safety.[106] Finally, the Australian researcher A. E. Ringwood has demonstrated that even wastes glazed with borsilicate can show increased water solubility within a brief period: this is detectable after only one month of storage in hot distilled water at 95°C.[209]

The salt deposits at Gorleben had in any case proven themselves to be worthless—they were too permeable to water.[125] The floor of the intermediate storage hall developed cracks in spite of all efforts, and buckled out. The nuclear-waste barrels began to rust, and were not, as claimed, absolutely break-proof. The dream of fail-safe final disposal seemed to be dead. No wonder Western nations are now discussing shipment of their nuclear wastes to China's Gobi desert. None of this is surprising. The well-known American pro-nuclear physicist, Alvin M. Weinberg, stated the following in 1984: "We nuclear engineers have been unable to convince the public that nuclear technology is a benign and acceptable technology. In 1977, 30% of the American public was opposed to nuclear power; today, 60% are."[98] Evidently, the honest research carried out by opponents,

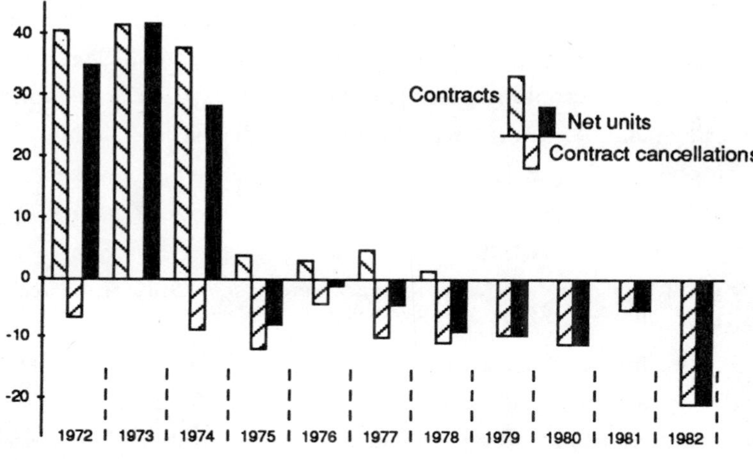

together with the Harrisburg accident, has been effective. Even in Switzerland, almost half the voters opposed nuclear power in a 1984 referendum.

Other countries also have strongly reduced their nuclear programs, including Spain and France. Mexico plans to build only two nuclear plants and five hydroelectric power plants.[213] The decision by France to call a temporary halt for reasons of cost, to any further construction of fast-breeder reactors caused much stir. Nonetheless, the 1200 megawatt "Superphenix" in Cray-Malville came on-line in 1986, and is to remain a prototype.[185] A decision on the continuation of the fast-breeder program will not be made until after 1987.

In Germany, fast-breeder technology is no longer being pursued. The nuclear power company of primary concern announced at the end of 1984 that it was not planning any second fast-breeder. Only the Kalkar prototype was to come on-line in 1985, after being under construction for 12 years; the costs rose within this and the preceding planning period from 500 million German Marks to 6.5 billion German Marks (the latter figure is approximately 4 billion U.S. dollars at current rates). The government of the state of North Rhine-Westphalia, where Kalkar is located, has stated that, contrary to predictions, the breeder would prove to be unusable.[185]

Originally, it was believed that "energy could be bred almost for free" by means of breeder technology. The fast-breeder reactors during the energy-production process, can transform the useless isotope uranium 238, which constitutes the major portion of natural uranium, into fissionable plutonium. In this way, a breeder comes to produce more nuclear fuel than it consumes. The plutonium then has to be removed from the fuel rods in a reprocessing plant. A plutonium economy, with all its incalculable technical and sociopolitical dangers, would be the consequence of these technological acrobatics. Just one millionth of a gram of plutonium can cause lung cancer. In September 1983, the United States Senate cut off the money flow for the prototype plant at Clinch River, Tennessee. Only small research projects are still being continued.

The dream of cheap nuclear power is dead for good. The respected Harvard Energy Report, published by the Western world's most famous business school, foretold this as early as 1979.[180] "For these reasons, we would be ill-advised to depend on nuclear power merely for the sake of more oil independence for the rest of the century."[190] The Swiss physicist, Ruggiero Schleicher, wrote the following in his much-discussed 1984 report "Nuclear Power—The Big Bust," published by the Swiss Energy Foundation (SES):

> Reequipment and maintenance of the aging reactors, as well as treatment and final disposal of radioactive wastes, will plague the nuclear industry for decades, even if there are no more significant orders for nuclear power plants. Here, too, high quality, reliability and good engineering work are of decisive importance for safety. A slowly dying industry, however, yearning for the good old days and suffering from loss of prestige will not attract talented, dedicated professionals. The most expensive solution, then, in the view of the Washington, DC-based Worldwatch Institute could therefore be to blunder along indecisively.... Military use—the bomb—which was at the beginning of the whole development, will remain the principal rationale for this technology.[155]

Coal-Fired Power Plants and Radioactivity

For many years, nuclear advocates have claimed that radiation exposure in the vicinity of coal-fired power plants is 100 times greater than those near nuclear power plants of the same size. This fairy tale has long-since been laid to rest. In 1977, UNSCEAR determined, while taking the entire fuel cycle into account, that the global collective dose of the population per megawatt of electrical energies is 375 times higher in the case of nuclear power than that of coal-generated power.[200] In 1975, Japanese scientists reported in *Health Physics* that the whole-body dose produced by Japanese coal-fired power plants at the point of maximum radioactive exposure was around 0.01 mrem/year.[124] This is 2000 times less than the maximum of 20 mrem/year allowed in the vicinity of Swiss nuclear power plants under the ALARA principle, or 100 times smaller than the supposed average exposure of 1 mrem/year in like areas. Radioactivity is something that can no longer be used to demonize coal-fired plants—except among poorly informed citizens.

III. FOREST DEATH AND RADIOACTIVITY

The New Dimensions of Forest Death

Forest death has taken on dramatic forms. Due to the rapid progression of damage, a study can hardly be completed on the subject before it becomes obsolete. Evergreens long ago ceased to be the only trees affected; beaches, ashes, oaks, and now even fruit trees like apple, pear and cherry are showing similar symptoms of damage. In the Swiss canton of Thurgau on Lake Constance, 30% of all fruit trees are affected. Foresters are afraid that grape-vines will be next.

The danger that is emerging is that the basic process of the life cycle of plants, animals and humans will be interrupted: photosynthesis is threatened. Photosynthesis is the means by which plants transform the light energy of the sun into stored energy, i.e., into the carbon compounds that make up the body of the plant. Plants build themselves up in a miraculous manner. As building materials, carbon is taken from the carbon dioxide in the air, and hydrogen is taken from the water which the roots absorb along with minerals. The chlorophyll of the leaves is the catalyst for this process; the sunlight is the source of energy. Without the plant matter which is created by photosynthesis, animals and humans would have no food. Science cannot preclude the possibility that ever more plants—including human food plants, will be detrimentally affected by forest death, and even die out.[39]

The question that should be asked is why plants react so much more sensitively to air pollution than do animals and humans. There is a fundamental difference between humans and plants. We need the air "only" as a provider of oxygen, to burn our food and provide us with energy. Plants, however, get almost all their building material, i.e., their food, in the form of carbon. It is contained in the air as carbon dioxide (CO_2), and is passed back to the plants by means of photosynthesis. However, a plant has to "breathe in" and process incomparably greater quantities of air than does a human, for air contains only 0.035% carbon dioxide, as opposed to 21% oxygen. For this purpose, the leaves and needles of plants are provided with a highly-developed ventilation system, so that enough CO_2 can be incorporated from the great dilution in the air. This air finds its way into the interior of plants and needles through very fine pores, the so-called stomata. A single oak or beach leaf has over half a million such openings.[39]

This intensive ventilation of plants explains their much greater sensitivity toward air pollution. Thus, the effects of air poisoning are noticed much earlier in plants than in animals or humans.

Background

Classic Forest Death

Forest death has occurred since the dawn of industrialization in the 19th century. Classic smog-damage due to sulphur dioxide (SO_2) and sulphuric acid (H_2SO_4), were already described by Haslehof and Lindau in 1903, and by Wieler in 1905.[115] Even then, it was demonstrated that the damage was manifested in lower growth rates (narrower growth rings in the trunks). "Today, it has long-since been proven that the width and structure of the growth rings are an expression of photosynthetic performance, i.e., of the vitality of the trees."[115]

This classic acute, or direct, damage to trees appears worldwide, mostly in the immediate or approximate vicinity of pollutant emitters such as fired power plants, heating plants, metal-processing plants, waste incinerators and the ceramics industry. However, in recent decades, the flue gases have been carried to ever more remote places, due to the current practice of building super-high smokestacks.[8]

New Types of Forest Decline

By contrast, a creeping, new kind of forest death began to appear, first in central European mountain ranges, such as the Harz along the old border between East and West Germany, and the Riesen and Erz mountains on the German-Czech border. During the 1970s, it seemed that only white firs were affected, and it was thought that familiar white fir disease had broken out. Ten years later, when the ailment began to spread to other species and move to areas of very clean air, far from emission sources, scientists were confronted with a great riddle.[113] Though a few scientists such as Professor Otto Kandler at the University of Munich continue to believe in a plague or epidemic theory, virtually no one takes it seriously.[49] Single pests such as bacteria or viruses, never attack several different species of plants. Also, the damage-pattern distributions that we see today clearly indicate emission-induced damage. The Forestry Testing and Research Center of the German state of Baden-Wurttemberg (FVA) considers the emission-related explanation to be verifiable. However, it adds, that this "does not establish in any way which pollutant or combination of substances, or concentration of substances is responsible."[106] Thus, we still know very little. As recently as 1983, Dr. F. H. Schwarzenbach, Deputy Director of the Swiss Federal Institute for Forestry Experiments (EAFV) in Birmensdorf, Switzerland, stated that "the majority of researchers agree that central European forest death must be considered a hitherto unfamiliar process of destruction."[113]

The most comprehensive compilation of information on forest death is probably the Special Report by the Environmental Council of the West German Federal Ministry of the Interior, published in 1983. It states:

> The new areas of declining forest are not comparable, or only partially comparable to the known central European areas of smog damage, such as the Ruhr Valley, etc. . . . Moreover, the type and extent of forest decline observed can no longer be explained on the basis of traditional forestry experience.[14]

Forest decline as remote from emission sources of pollutants is particularly puzzling—although the connection with emissions appears certain. As the Council states:

The virtually simultaneous occurrence of this damage leads to the suspicion of a sudden, almost shock-like effect of a new and unknown source of damage—also known as Factor X—on which all of it would be blamed.[14]

Recommendation of Schütt Concerning Radioactivity

Pro-nuclear scientists particularly, do not like to hear the term "Factor X," since it might refer to artificial radiation. If such a connection were proven, the rapid end of nuclear power would be inevitable. The famous forestry scientist Professor P. Schütt, who holds the chair in forest biology at the University of Munich, courageously recommended in 1983:

> . . . . that radioactive emissions also be investigated seriously and comprehensively as a possible cause of the syndrome [of multilayered disease phenomena, in this case specifically forest death].[98]

Indeed, the increase in the contamination of our environment by fission products, must not be overlooked or trivialized by often specious comparisons with natural radiation. "What we are doing today is like comparing the weight of bread with that of potassium cyanide—which is valid, as far as the total mass is concerned—in order to prove the safety of potassium cyanide."[135]

The Effect of Electromagnetic Forces

Radar facilities, radio and television transmitters, etc., may also have a connection with forest death, due to the radio and microwaves they emit. Schutt points out that research programs on the effect of electromagnetic waves barely exist, and adds: "Due to their connection with military and economic interests, this question too would, analogously to that of radioactivity, require additional efforts in order to reach objective clarification."[110]

Not even the effect of electromagnetic radiation on humans and animals is completely known. Living tissue that is heated by electromagnetic waves may be accompanied by effects other than an ordinary increase in temperature.[104] Thus, for instance, there are reports of changes in cell-membrane permeability (Berteaud).[104] However, according to the interim report of the "Manto" research project undertaken by the Federal Institute of Technology in Zurich, published in December 1984, the existence of non-thermal effects in the radio wave-length range of 30 kHz to 30 GHz, is highly controversial.[104]

All effects appear to be predominantly frequency-dependent.[104] For instance, the human body has a pronounced resonance point for the frequency of 17 MHz of a ham radio band.[59] (If the wave length has the same order of magnitude as the dimensions of the biological system concerned, resonance phenomena will appear. In such an instance, incoming wave-radiation will be absorbed more strongly, leading to increased heating. Resonance effects may increase thermal load by factors up to and even greater than 100.) The 900 MHz band, authorized in Switzerland for cordless telephones, operates at a wavelength for which the body has a particularly high absorption rate, and for which there is resonance for the head.[59] What is ominous is that there are large individual sensitivity differences.[59, 104]

Temperature increase has so far been the central factor in establishing maximum limits. In Western nations, American safety standards are generally taken as guidelines.[59] In the range of 30-10000 MHz, the resonance phenomena are taken into account in those terms, and a maximum power density of 1 mW/cm^2 (1/1000 watt per square centimeter; 6 cm^2 = approx. 1 sq. inch) is established.[59] Yet because the opinions of experts regarding possible effects vary widely, safety regulations in individual countries may vary by factors of 1000. This was discussed in a presentation by Chen et al. at the 1985 Symposium of Electromagnetic Tolerance at the Federal Institute of Technology in Zurich.[104]

This report also deals with an exhaustive epidemiological study of 423 persons who were occupationally exposed to electromagnetic waves over the course of three to nine years, and more. These persons showed significantly increased problems of the nervous and digestive systems, the blood and circulatory system, the heart, as well as cataracts, as compared with a non-exposed control group. The operational frequencies were between 2 and 9 GHz and 140-180 MHz. The power density around the working area was less than 100°microgramW/cm^2 (=0.1 mW/cm^2), or less than one hundred millionth of a watt.

The "Manto" interim report draws the following conclusions: "The discovery that in the range between 30 kHz and 30 GHz, below the power density that leads to discernible heating of tissue, no biological effects are ascertainable may now be considered certain, the large number of publications to the contrary notwithstanding."[59]

Given this critical situation, it is good news that Dr. F. Schwarzenbach, Deputy Director of the Swiss Federal Institute for Forestry Experiments, proposes an investigation of the health of forests around various broadcast facilities, with the aid of infrared aerial images.[73, 145] The high-frequency penetration depth is fully 30 cm (1") at 100 MHz, and still 3 cm (1 1/4") at 1000 MHz. Plant organs such as needles, leaves and cambium are within the range of total penetration. This is why the sap-flow of a tree under the bark acts like a radio or TV antenna—or as a lightning rod.[41]

The Swiss Federal Ministry for Environmental Protection states: "The non-thermal effects of radio and microwaves of low power density have not yet been conclusively demonstrated for higher plants. They seem to occur only in a few narrowly-limited frequency ranges. However, very little is yet known regarding the frequency ranges involved, or the associated threshold values."[146]

Intensive research for conclusive clarification must be sought in this area as well—in spite of all resistance. According to the "Manto" interim report, no studies can be found concerning the effect of radio waves on plants.[59]

So Far, Only Classic Pollutants Are Taken Into Account

The German Special Report takes account of only the following pollutants: combustion of all kinds of fossil fuels in power plants, industry and household heating, and motor vehicles.[2] Only the classic pollutants are considered:

- sulphur dioxide (SO_2). SO_2 occurs during the combustion of fossil fuels, particularly oil and coal.
- oxides of nitrogen (NOx). (This is a collective term for several oxides of nitrogen, particularly NO and NO_2; these are gases released during high-temperature combustion, such as in internal combustion engines.
- hydrocarbons. These are compounds of carbon and hydrogen.

- aerosols with heavy metals such as lead, cadmium, nickel, thallium, copper, zinc and mercury. Aerosols are tiny particles suspended in the air.
- photo-oxidants, such as ozone, hydrogen peroxide and peroxiacetyl nitrate (PAN). These are gases toxic to plants which occur as the result of the effect of sunlight on NOx's and hydrocarbons.

With this brief list, radioactivity is completely eliminated as a potential source of damage—a very unscientific examination.[14]

I was the first to disseminate Schütt's research to a broader audience in articles published in the *Basler Zeitung* of November 5-December 4, 1983.[33, 34, 35, 81] He was able to base his information on studies, in some cases unpublished, by the German physician Dr. K. J. Seelig, the Swiss engineer P. Soom, and the practical experience of Professor G. Reichelt.[78, 79, 80, 85, 91, 92, 98]

Nuclear power advocates reacted rapidly. They denied any connection between radioactivity and forest death, and countered Schütt by labelling his research insubstantial.[1, 15, 16]

Growth Retardation Throughout the Northern Hemisphere

I also found two very interesting studies by Dr. F. H. Schweingruber of the Swiss Federal Institute for Forestry Experiments in Birmensdorf. On the basis of drilling samples, it is assumed that the "decisive" physiological damage resulting in current forest death must have begun during the 1950s. This is depicted in a reduction of density and width of tree rings, and in reduced growth, which is true of the entire Northern hemisphere, and even the Himalayas.[114, 115, 116] Comparable phenomena in dated historical and prehistorical spruce trunks do not exist, according to Dr. Schweingruber.[114, 116] The Swiss Federal Office on Environmental Protection notes: "This is without parallel in the history of forests."[23]

Of course, temporary retardation of growth for various reasons (climate, pests, local air pollution) has been known for centuries, but has never gone beyond a regional context to a worldwide phenomenon, such as exists today.

As the following figure shows, the closer growth rings are spaced, the less growth the tree shows:

Trunk cross-section of a white fir felled in 1982. It shows extreme retardation of growth since 1958, and virtually no growth since 1970.

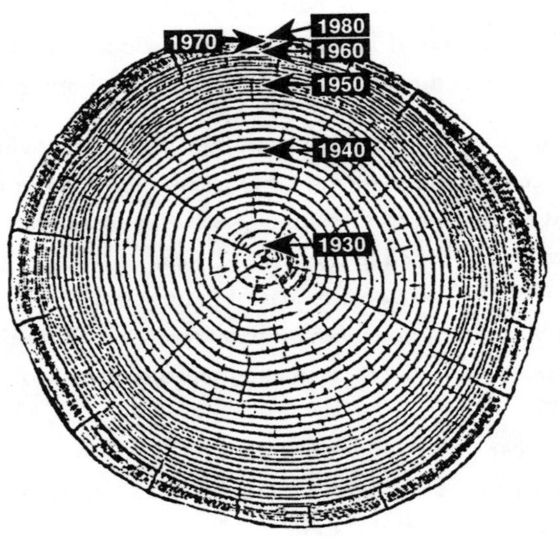

Such initially damaged, physiologically weakened trees—which do not necessarily show any outward decline, as in the case of firs—may later suffer varying kinds of growth retardation. Or, they may abruptly cease to grow altogether, possibly from further harmful effects.[110, 116] Then, following a delay of years or decades after the decisive initial damage, the trees will die.

Neither aging, inappropriate location, poor forest management nor climate can be considered as the possible sole cause of damage.[115] Forest death is rampant on poor soil and good soil, acidic and alkaline, dry and damp—all in equal intensity, according to Schütt.[116] The primary cause cannot possibly be lack of nutrients, wind, snow, fungi, bacteria or insects, either.[23] "Some overarching effect is causing the damage," says Schweingruber.[116]

The growth-ring profile of a tree shows exactly what effects the tree has experienced, both in terms of time and seriousness. In this way, the life and health history of a tree can be traced. On the other hand, its external appearance—crown, branches, leaves—provides only a fleeting image. Drilling samples of growth-rings should be an indispensable part of the investigation of forest death.

Hypotheses and Causes

During the 1950s and 1960s, there must have been a global wave of air pollution which caused the initial damage. It was certainly not our cars with NOx emissions and photosmog that were to blame, nor is it likely that SO_2 was solely responsible. Like all classic pollutants, SO_2 and NOx's are rinsed out of the air fairly rapidly by the rain. However, the Northern Hemisphere contains the world's industrial belt, the so-called Ferrel Cell between the 30th and 60th parallels of northern latitude. Historically, only minimal and relatively slow air-mixing occurs between this and neighboring zones.[91, 111, 131] However, this zone also contains the most nuclear power plants—over 300 of them—and almost all nuclear fuel processing centers; these are the world's most important radioactive air polluters. Also, the vast majority of nuclear weapons tests, with their dangerous radioactive fallout, occurred in the Northern Hemisphere. By contrast to SO_2, NOx's and hydrocarbons, certain artificial radioactive fission products are not even flushed out of the atmosphere; these include the radioactive inert gas krypton 85 (half-life 10.7 years) and radiocarbon, or carbon 14, with a half-life of 5,730 years. These products disperse and accumulate globally.

Certain hypotheses have already been suggested regarding the cause of forest death. First, it was thought to be the SO_2, then acid rain, then ozone and now, since 1983, scientists have suggested stress.[109] No one today doubts that some kind of combination or complex effect is the cause. It is unscientific to exclude artificial radioactivity, which is a truly global form of biosphere pollution. There does seem to be a basic harmful effect which is causing primary physiological damage.

As if without a clue, the then West German government's Council on Environmental Question stated:

> According to existing findings we have been unable to identify any single air pollutant as the sole cause. There are many indica-

tions that the damage is caused by the interaction of many simultaneously or sequentially attacking damage factors; i.e., it is a synergistic effect.[14]

S. McLaughlin of the Oak Ridge National Laboratory in the U.S. and O. U. Braker of the Swiss Federal Institute for Forestry Experiments have written that in spite of various hypotheses, no proof has yet been presented that any single harmful agent is the main cause of damage.[58] Atmospheric pollution as the consequence of large-scale combustion of fossil fuels over the course of the last three decades, including SO_2, ozone, acid rain and trace metals, were all cited as possible causal factors.

The Automobile Industry Association of Switzerland (AGVS) is quite right to reject measures against automobiles. The official thesis on forest death has no serious basis, and is based on a compilation of facts, half-truths, simplifications, unchecked hypotheses and ideologies with more support in politics than in science, according to the AGVS.[72]

The Acid Rain Hypothesis

The acid rain hypothesis is based on the assumption that sulphuric, nitric, hydrochloric and carbonic acids lead to chemical reactions in the soil. This liberates plant-toxic aluminum and manganese ions, damaging to root-hairs. Also, the effects on the above-ground portions of the trees are taken into account.[109]

However, if SO_2 (sulphur dioxide) is dissolved in water in the lab, nearly the only thing that is produced is sulphurous acid ($H_2\ SO_3$), which is much weaker than sulphuric acid ($H_2\ SO_4$). In order to produce the very acidic sulphuric acid, it is first necessary to oxidize the sulphur dioxide (SO_2) to sulphur trioxide (SO_3). The mechanics of this process are not really known.[14, 17] It is thought that photo-oxidants such as ozone and hydrogen peroxide (generated by the effect of sunlight on NOx's and hydrocarbons), act as catalysts or promoters of the reaction.[14, 17] However, artificial radioactive substances may have the same effect. Artificial radiation produces radiant energy, and can directly oxidize SO_2 to SO_3, or form ozone from atmospheric oxygen. Radiolysis in water can form hydrogen peroxide directly, and may even produce NOx from atmospheric nitrogen.

Pro-nuclear scientists cannot refute these physically well-proven facts. However, they can use purely theoretical models to claim that in normal operation, it is quantitatively impossible for nuclear facilities to affect forest death significantly.[16, 53, 61, 81] A report of the Institute for Energy and Environmental Research (IFEU) in Heidelberg, Germany, came to the same conclusion.[122] It should, however, be noted that this study was made prior to Reichelt's epoch-making discovery.

Moreover, independent scientists obtain different quantitative results from their theoretical calculations.[52, 62] For example, Messerschmidt recommends systematic measurements in the immediate vicinities and regions of nuclear power plants, for substances such as ozone.[65] Theoretical calculations can never replace direct measurements and comparisons. The Federal Office for Environmental Protection in Bern believes that theoretical considerations in which natural and artificial radiation are regarded as similar in their properties, are inadequate. Therefore, it is not justifiable to conclude that radioactive emissions from nuclear power plants can under no circumstances be the cause of forest death.[145]

At a 1975 meeting of the International Atomic Energy Organization, a number of very interesting correlations involving radioactivity were presented by the scientist K. C. Vohra of the Bhabha Atomic Research Center at Trombay, India.[134] Within a 2-km circle, there is a 40-MW research reactor (Cirus) and a conventional coal and oil-fired power plant. Vohra began with the assumption that so-called condensation nuclei are constantly forming in our atmosphere by means of chemical reactions of various substances. He was able to determine by experiment that this condensation-nucleus formation increased somewhat in the SO_2-rich exhaust gases under the effect of sunlight or cosmic radiation. However, when radioactive gases were released from the nuclear power plant, condensation-nucleus formation increased rapidly. The SO_2 was quickly oxidized to SO_3, which in turn leads to sulphuric acid, which also forms condensation nuclei much more readily.

The graph below shows Vohra's crucial results:[134]

"These experiments have clearly shown that the combined effect of SO_2 and radioactivity leads to heavier condensation-nucleus formation than does normal photo-oxidation in sunlight," says Vohra, i.e., sulphuric acid is formed much more quickly and more completely than by means of

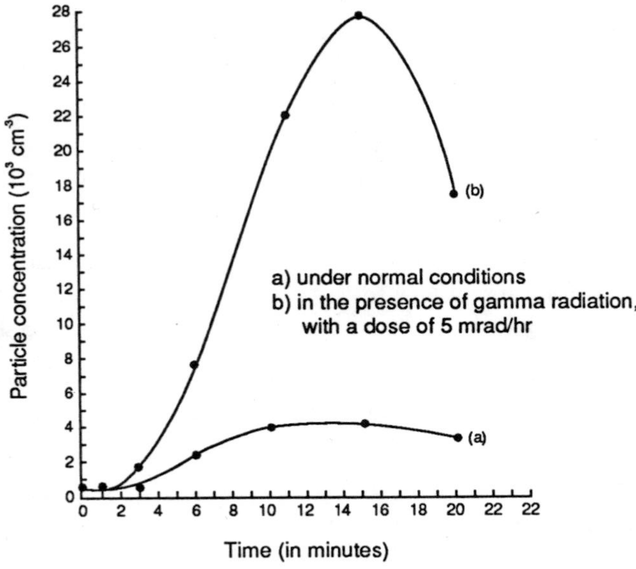

photo-oxidation. As early as 1973, at a symposium in Vienna, Vohra had said:

> It should be noted that the increased use of fossil fuels for power generation is leading to the release of ever greater quantities of SO_2, while the marked increase in the number of nuclear power stations is causing higher doses of radioactive emissions into the atmosphere. Studies of the combined effect on the atmosphere of these two types of emissions therefore deserve the greatest attention.[46]

It is significant that these warnings have simply been "forgotten." A broad field of research is lying fallow. In 1981, Vohra and Subba Rhamu demonstrated that particle formation (condensation-nucleus formation) can occur explosively if additional reagents and especially aerosols or dampness are added.[46, 86, 102]

A European measurement network, the European Atmospheric

Chemistry Network for the measurement of the pH of rain, has existed only since 1950.[17] "pH" is the expression of the acidity of a solution, extending from pH 1 (extremely acidic) through pH 7 (neutral) to pH 14 (extremely alkaline or basic). A change of pH by a value of one represents a change in the degree of acidity by a factor of 10.

The table below shows the spotty results published; measurements have been taken only since 1956. Nonetheless, the increasing degree of acidity (falling pH) had been clearly visible during the past decades.

Country	Old pH:	New pH:	
Southern Norway	5.5–5.5 (1956)	4.7	(1977)
Southern Sweden	5.5–6.0 (1956)	4.3	(1978)
Italy	—	4.3–6.5	(1981)
Black Forest	—	4.25	(1972)
England	4.5–6.0 (1956)	4.1–4.4	(1978)
West Germany	—	3.97	(1979–81)

By 1984, in broad areas of Europe and North America, precipitation had values as low as 4-4.5. In the Swiss midlands, the average is 4.5.[23]

We don't even know for sure what the pH of rainwater was prior to the nuclear age. Theoretically, pure rainwater should have a pH of 5.6.[23] However, in 200-year-old Greenland ice, pH's of between 6 and 7.6 have been found. There, however, the displacement and transformation of material in the ice cannot be excluded. In any case, Schütt has stated that "a conclusive judgment regarding the pH drop of rainwater is hardly possible at present."[110]

Unfortunately, the European Community does not seem to be able to think globally: it is silent on the question of Asian conditions.[14] This is why we must depend on other sources. Since 1972, there has been a measurement network for the entire Northern Hemisphere, the "World Meteorological Organization" (WMO). H. W. Georgii and others have illustrated the average pH-precipitation data of the entire hemisphere for 1979.

The industrialized centers of central Europe and North America can be clearly seen with low pH values, at the same time large parts of Asia still show values of 6-7. However, since growth retardation of trees has occurred during the 1950s and 1960s throughout the Northern Hemisphere, and even in the Himalayas of Nepal, it is no longer possible to blame only the classic pollutants—SO_2, NOx's and hydrocarbons. We must also consider the truly global polluter, artificial radioactivity, whatever its operative mechanisms may be. These are hard statements that we must carefully ponder; but we cannot simply dismiss the possibility of synergistic or amplifying effects of artificial radioactivity interacting with classic pollutants in highly industrialized countries.

There is hardly anything more simple than a pH measurement. Why were systematic measurements not undertaken earlier? Perhaps there was no interest in an early warning. If something has not been measured, it is not there. It is disturbing to read in the West German Energy Report of March 16, 1984, published in Bonn, that the origin of acid rain can be explained by the fact that it has been around for a century. That is true—even with pH values as low as four—though only in the immediate vicinity of pollution emitters, not in clean-air areas.[23]

Why rain has become more and more acidic since the onset of the nuclear age is an open question.[91, 98] The emission of SO_2 in West Germany remained constant, for all practical purposes, between 1966 and 1980, and the NOx emissions that result in nitric acid pollution rose by only one third between 1966 and 1978.[13, 14] In Switzerland, the 1984 emissions of sulphur dioxide were actually lower than those of 1955; however, NOx emissions

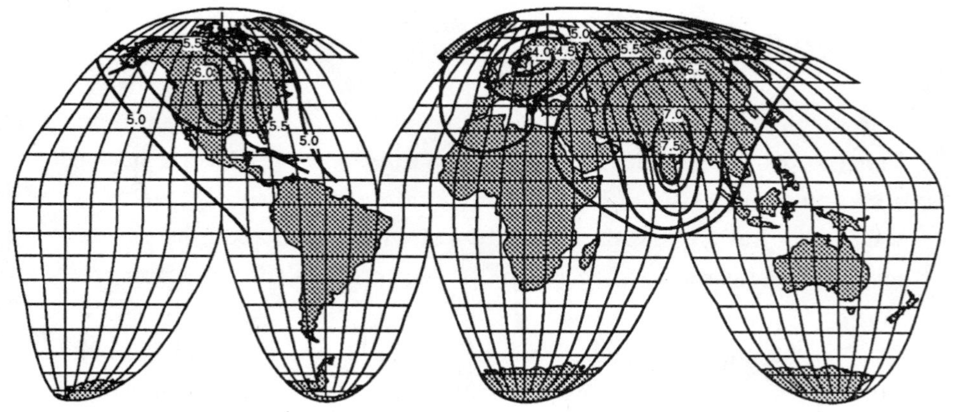

have been some two to three times as high in 1975 as they were in 1964.[23, 97] A study by the Lausanne Technical College at the Col du Gnifetti Glacier, 4,450 meters (13,350 ft.) high in the Alps near Monte Rosa, indicated that the degree of acidity had risen from pH 6 to pH 5 between 1965 and 1979, after having remained more or less constant at the former figure since 1916.[124] This decline in the pH value by one, corresponds to an increase in acidity by a factor of 10. There is no direct relationship between the quantities of SO_2 and NOx's emitted and the precipitation of sulphuric and nitric acids in the rain, snow or fog droplets.[96]

According to J. Fuhrer of the Institute for Plant Physiology at the University of Bern, the composition of acid rain and acid deposits are determined by photochemical and wash-out processes, as well as by air exchanges in the atmospheric interface layers. Many of these processes are not yet fully understood.[29]

In addition, a portion of the pollution is deposited dry—without having affected the rain—especially NOx.[23] Given equally high emissions of SO_2 and NOx, a rainwater droplet will contain twice as much sulphuric acid as nitric acid.[23]

Furthermore, the composition of fog droplets is insufficiently known. It is, however, known that fog acidity may be considerably higher than that of rain.[28] Thus, according to the Swiss Federal Institute for the Protection of Water, the accumulation of pollutants in fog is often 10 to 100 times greater than in rainwater. This is because there is less liquid in a cubic meter of fog than in the same volume of a rain cloud.[22, 121] Unfortunately, this study does not take radioactive substances into account. They too are pollutants.

It is unscientific to deny *a priori* the participation of artificial radioactivity in forest death and to, in effect, choke off research. Reality may differ from theoretical model calculations, as Vohra has shown with his experiments.

SO_2, NOx, and hydrocarbons are also emitted from natural sources. Volcanic eruptions, fires, lightning, ocean and sea spray, the metabolism of organisms, and natural decomposition processes are in the main the originators. Lightning produces the largest share of natural NOx, and a single volcanic eruption can spew out more SO_2 than industry emits in an entire year.[14] The following information from Ullmann's Encyclopedia of Chemistry, 1981, shows the global activity, as opposed to the naturally-produced share:

SO_2 (sulphur dioxide	13.3%
NO_2 (oxides of nitrogen considered NO_2)	6.5%[128]

By contrast, the Swiss Federal Office for Environmental Protection reported that the anthropogenic share of SO_2 and NOx emissions in the industrialized countries of Europe was approximately 95%. Globally, natural and anthropogenic shares of SO_2 and NOx emissions are equally great.[14, 23] According to Hornbeck (1981), the natural share of overall emissions is controversial in the literature as well.[14, 42] Schütt, too, refers to widely varying estimates for SO_2 emissions.[110] Now that forest death has spread across the entire Northern hemisphere, global thinking is encouraged. In 1983, a report by the Brookhaven National Laboratory published in *Science* magazine states that auto traffic in the eastern U.S.A. contributes less than 14% of the strongly acidic fractions of acid rain. For this reason, stricter traffic laws would not cause any improvement.[38] Continual measurements over the past 20 years have been unable to draw any direct connection between NOx emissions from auto traffic and NO_3 ions, the nitrate ions of nitric acid, in the acid rain.[89]

Americans were spared until 1984, when considerable forest death was noticed in the Appalachians—i.e., in the east. But even in the 1950s and 1960s, growth retardation had been taking place there.[47, 74] Additionally, since 1970, 15 major nuclear power plants have gone on line in Alabama and Tennessee, west or upwind from the mountains.[196] Here, as well as in the Carolinas to the east, downwind from the Savannah River plutonium and tritium production facility, the greatest damage has been reported. However, no new coal-fired power plants have been added during this period.[118] On Mt. Mitchell, high ozone concentrations were suddenly measured.[74] Yet, the U.S.A. has long had catalytic converters and lead-free gasoline.

Japan, too, has taken all these precautions since the 1970s, as well as filter devices in heating plants; yet, there too, forest death exists. In 1920, the Meiji Shrine in central Tokyo, contained 365 native tree species adaptable to the location. In 1984, only 247 remained: spruces, cedars and other evergreens had been decimated or eliminated altogether.[70] The Japanese islands are ecologically favored by abundant rain storms and strong westerly winds sweep the wastes from factories and exhausts into the Pacific.[2]

There are mysterious reports, too, from Sweden.[20] Forest damage is well known in the south and the west of Sweden, and is supposedly caused by polluted air blowing in from central Europe and Britain. Recently, however, Swedish foresters have discovered acutely sick trees on a large scale, in northern Sweden. Even in the west and south, the value for oxide of sulphur and nitrogen is only half that of those common in central Europe; in northern Sweden, they are only 5-10%. The foresters have no idea as to the cause of this new phenomena.

The Ozone Hypothesis

Ozone, like hydrogen peroxide and peroxiacetyl nitrate, is a photo-oxidant. It is formed under the effect of the high-energy ultra-violet components of sunlight as a result of the photochemical transformations of nitrogen dioxide and hydrocarbons.[14] Schütt writes: "According to the ozone hypothesis, the participation of ozone (O_3) is decisive. Even under the climatic conditions prevailing in central Europe, enough ozone is said to be created due to the photo-oxidation of the oxides of nitrogen released from internal combustion engines so that quantities toxic for forest trees can occur at any time. First of all, this leads to direct damage to leaves, and secondly, it is said, the permeability of the cell membranes to acid precipitation is increased, and nutrients can be flushed out. An example is the loss of magnesium from spruce and fir needles, which soil analysts have ascertained."[109]

Under this theory, auto exhausts (considered the primary ozone sources) can be transported over 100 km through the atmosphere, until a high ozone concentration develops at the remote location of its origin.[14] This is the basis for demands by Germany to counteract forest death with freeway speed limits. By contrast, Dr. P. Jakober, professor of ecology at the School of Engineering in Burgdorf, Switzerland, notes that a reduction of the speed limit would have a ridiculously minimal effect.[71]

It is not credible that the ozone generated by automobile traffic during the 1950s and 1960s caused the global growth retardation of trees and laid the basis for the death of the forest. On the other hand, damage from ozone generated by transformed auto exhausts is indeed possible on a regional level. In the end, former West German Economics Minister

Martin Bangemann opposed speed limits, opining that he was waiting for proof that auto exhausts affect forest death.[4]

A number of factors weigh against the ozone hypothesis. For instance, forest death in Switzerland began on a large scale in 1981, though the mean ozone content of the air had risen only minimally during the period between 1980 and 1983. In addition, the mean SO_2 and NO_2 content had remained constant during the same period.[12] According to Schutt, the high point of spruce damage occurs during the cold half of the year, when harmful ozone concentrations hardly form.[110] There is also intensive forest decline along freeways and in city centers, where ozone plays a minor role. Moreover, the damage is not associated solely with chlorophyll reduction and yellowing of leaves, apparently typical for ozone poisoning. Deciduous trees tend to be more sensitive to ozone than evergreens.[110] Observations by Prinz et al. of West German forests around Hamburg and near Hills, Hillenbach and the Egge Mountains in North Rhine-Westphalia go so far as to dismiss any significant involvement of ozone in forest decline.[43] Likewise, on the basis of biochemical and physiological data, Huttermann believes that photo-oxidants such as ozone can have no decisive part in the forest death in the Egge Mountains.[43]

The West German Interior Ministry's Environmental Council is clearly saying "may be" rather than "is" in the following statement:

> Photo-oxidants may be of significance for emitter-remote forest decline which has until recently been strongly underestimated.[14]

Moreover, ozone can be produced in the lower atmosphere by radioactivity, utterly independent of sunlight. Thus, such radioactive inert gases as krypton 85 and xenon 133, which are released unchecked in all atomic fission processes, are known as effective ozone-generators. This is particularly true of long-lived krypton 85, which is dispersing and accumulating worldwide.

As mentioned above, nuclear advocates do not argue with the validity of these reactions, but do claim that they are too rare to be of significance. At the same time peak exposures, which are of great importance are paid much too little attention. Reichelt does refer to calculations made by Kollert when indicating that, while artificial ionization density in nuclear plant vicinities might be two or three orders of magnitude below average natural radiation, it can for certain periods of time rise to two to five orders of magnitude above that level; the unmixed package of air containing radionuclides can then be transported across great distances.[88, 52] According to Metznert, such pollution plumes can be shown to extend for over 50, and even over 100 miles.[147] If the ozone production per square kilometer of a reprocessing plant is taken and compared with that from natural sources per square kilometer, a local ozone concentration increase by several orders of magnitude can result.[147]

The continual increase in ozone concentration at 900 meter altitudes in the Northern Hemisphere is notable (Johnston, 1984), as is the increase in ozone pollution during the cold half of the year, which cannot be explained merely as photosmog.[48, 129] Reichelt therefore points out that the smog theory should be reexamined in terms of cool-to-moderate climates, particularly in regard to generation solely by short-wave light.[88]

Let it be stated clearly: there is no argument over whether acid rain or all ozone is generated by artificial radiation only; this would be disproved merely by the seasonal variation in the ozone level. Nonetheless, we should recall once again the experiments performed by Vohra, which came upon surprising and partially unexplained discoveries. Atmospheric physics and atmospheric chemistry have some catching up to do. They will not get very far by theorizing; targeted, comprehensive research that Vohra had already recommended in 1975 will be needed.

Of course, nuclear advocates are uninterested in such studies, since information detrimental to them could turn up. It might be reported that the dispersion models for emissions from the smokestacks of nuclear facilities are not reliable. Many of the advocates assume a comfortable, relatively rapid dispersion—and hence dilution—in the atmosphere, which obviously is not always the case.

In this connection, it is noteworthy that the European Community has located the European Research Center for Air Purity Measures at the Karlsruhe Nuclear Research Center in Germany. Germany's own forest death research—at least that which is conducted by the Federal Ministry of Research—is carried out and coordinated by the Nuclear Research Center.[133] No wonder, the inclusion of radioactivity into research on forest death meets with strong resistance.

The Stress Hypothesis

Air pollution, extreme weather conditions and pathogens produce stress on plants. It affects the hormonal balance of plants. "Consequences observed from such disturbances include premature aging, reduced stomatic width, premature leaf-fall and growth retardation."[110]

According to Schütt, the stress hypothesis is based on the fact that for decades, small concentrations of pollutants in the air have been causing the continual disturbance of photosynthesis. This pollution causes a continual deficit in carbohydrates produced, which in turn leads to reduced vitality, disruption in root and leaf renewal, and hence, greater susceptibility to secondary damage.[109, 110] Thus, in the stress hypothesis, all known facts about forest death form part of a mosaic-like picture.

Schütt, however, does cite arguments against the stress hypothesis. He brings up the question of why the pollutants do not take effect sooner, at least locally, since the small concentrations of pollutant-combinations have been present for some time.[110] Thus, the sudden appearance of forest death remains unexplained.

Symptoms of Forest Damage

We will not discuss the new types of symptoms; for this, we refer the reader to Schütt's excellent book *The Forest is Dying of Stress.*[110] Forest decline is now found everywhere; in gardens, parks and orchards. Obvious growth-ring fluctuations and other growth abnormalities are being discovered in firs. "The typical symptoms of damage include general thinning of the crown, a discoloration of the old needles of spruces and firs, "stork's nest"-type formations in firs, weak foliage on individual branches, and premature fall coloring in beach trees."[24] There are also characteristic twig and leaf deformations.

In spruces, it was found that sickly stands are blossoming strongly and put out larger-than-usual quantities of cones and seeds every year. Young spruces too, are blossoming, a previously uncommon phenomenon. Biology teaches us that organisms can call upon great reserves for reproduction if their environmental conditions deteriorate dramatically.[110] This seems to be confirmed now by the recent forest death. For instance, beech trees have been uncommonly fruitful in recent years. Apple, pear and cherry trees are yielding rich harvests, even though they are obviously sick. No one knows where this will lead.

Operation Noah's Ark

An ominous development is the impairment of the sexual reproduction of trees. The maintenance of genetic diversity and adaptability is an indispensable precondition.[110] The forests have a genetic heritage shaped by generations of natural selection. Species adapted to local and regional environmental conditions can be lost irretrievably. Only the appropriate seeds will make reforestation at all possible. For this reason, the seeds of valuable stands are already being collected in Switzerland, in the so-called "Operation Noah's Ark" of the Federal Institute for Forestry Experiments in Birmensdorf.

Three Particularly Dangerous Radionuclides

Radiocarbon Carbon 14

Radioactive carbon, or carbon 14, is a beta-emitter and is generated either naturally by cosmic rays, or artificially by nuclear weapons tests and nuclear facilities. When released, it is distributed globally in the air as carbon dioxide. Its half-life of 5,730 years, and the concomitant long-term genetic and somatic damage, including cancer, are particularly ominous.[72, 123] Carbon 14 is harmful beyond its radiological effect. In decomposing to form nitrogen, it can destroy genetically important molecules. Carbon 14 is not retained by the filters in the ventilation systems of nuclear power plants; in official emission reports, this beta-emitter had been simply ignored until 1973.[7, 77]

Early UNSCEAR reports are a gold-mine of information on the pollution of our environment by A-bomb tests. An interesting figure, for instance, is contained in the 1969 edition.[34, 132] As shown below, by 1963, carbon 14 had increased by 100% over its natural content in the northern troposphere—up to eight to ten kilometers (See Figure below); in 1984, the figure was still up by 25%.[55, 61]

The figure also shows that since the beginning of the nuclear age,

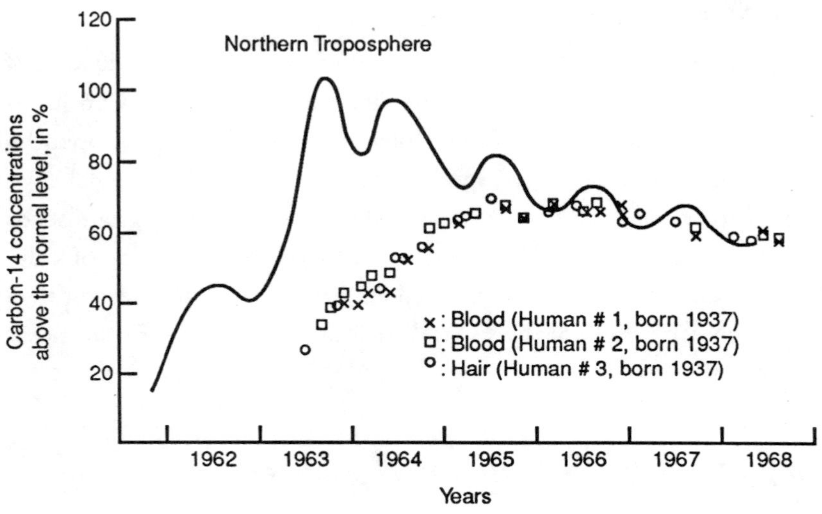

the carbon 14 content of human blood and hair has increased in direct proportion with the carbon 14 content of the troposphere. Unfortunately, corresponding data is not available for plants or trees. Yet, in the vicinity of various nuclear facilities, significantly increased Carbon 14 concentrations were ascertained in leaves and bark.[55, 57, 60, 76, 95, 145] An American study recommends that no agriculture be carried out in the vicinity of nuclear power plants.[26]

Usually, when air pollution is mentioned, it is taken to refer to the classic pollutants already discussed, and the much more serious contamination of our biosphere with artificial radioactivity is often completely overlooked, even by scientists. In 1972, the American National Academy of Sciences stated in BEIR report III that with the development of atomic energy, the biosphere would inevitably be subjected to increasing radiation exposure.

According to UNSCEAR 1982, the collective dose of the population due to artificial radiation increased by a factor of 100 between 1960 and 1980.[100] It is then noted that these doses are only averages, not necessarily the actual radiation exposure for an individual.

In contrast to classic pollutants, carbon 14 cannot be easily flushed out of the air through precipitation (rain or snow), any more than can carbon dioxide containing normal carbon 12. Therefore, an accumulation in the air is inevitable (see the above graph).

Seelig and Soom (the latter in his memorandum on forest death) have referred to the work of the tree-dating expert E. Hollstein of Trier, Germany.[40, 98] By testing antique wood carvings dating from 70 B.C. to the present, Hollstein discovered a connection between long-term growth fluctuation of oaks in central Europe and the global oscillation of carbon 14 content of the biosphere. An increase in radioactivity of only one percent was found to correspond to long-term damage to tree growth of approximately 18%.[40]

These fluctuations of the carbon 14 content in the growth rings of trees can be traced back to sun-spot activity, as expressed in the 11-year sun-spot cycle, pointed out by the United Nations as early as 1962.[115, 130] The production rate of carbon 14 in the atmosphere is dependent on cosmic radiation. The latter is subject to the relatively short-term effect of solar activity—the 11-year sun-spot cycle. The carbon 14 is absorbed by plants by photosynthesis of carbon dioxide, and built into the growth rings and other parts of the plant. In this way, the carbon 14 content of the growth rings corresponds to the current condition of the atmosphere, in terms of carbon 14.[115] Any synergetic effects which may occur will significantly complicate this picture, however. It is unknown whether growth retardation at the time of increased carbon 14 content is indirectly due to climatic variations of changing solar activity, or directly due to the harmful effects, as discussed earlier, of carbon 14 content in the growth rings. The Swiss Federal Environmental Protection Office in Bern also takes note of the incorporation of carbon 14 by trees.[145] According to Stuiver and Quai, however, a relationship between solar activity and climate is difficult to prove; Süss will not exclude it as a possibility.[103, 120]

Thorough research of this question is required. Increased incorporation of carbon 14 definitely leads to damage to plants, as Reichelt supports.[85] In 1982, the carbon 14 content of tree leaves was still 25% higher than the natural level prior to the atomic age.[56] Most of this increase is due to the A-bomb fallout, and is decreasing by about two percent per year.

Significantly, this unprecedented increase of carbon 14 in the

Northern Hemisphere's atmosphere, together with that of other fission products like tritium and krypton 85, parallels an unprecedented global retardation of tree growth, reaching all the way to the Himalayas. No climatic fluctuations are responsible for this.[14, 23, 115]

Tritium (or Radioactive Hydrogen)

Tritium is formed in small quantities naturally, by cosmic radiation in the upper atmosphere, but also artificially, by nuclear fission. It was not discovered until 1950, and for a long time its harmfulness was underestimated. "The literature on tritium has become overwhelming," says F. N. Flakus of the International Atomic Energy Agency (IAEA).[27] Tritium is a beta-emitter with a half-life of 12.3 years.

Tritium can accumulate, due both to its long storage time in ecological systems, and due to its so-called isotopic effect in organic material, for instance the food chain. Literature on this form of radioactivity is extensive (e.g.,[10, 44, 62, 90, 92, 127, 139, 141]).

Moreover, Seelig has pointed out that tritium may be incorporated into organic material more easily, and in greater quantities (by a factor of 10 to 100), when in the presence of iodine, certain heavy metals, microwaves or "silent discharge."[92, 93, 139] It is urgent that the possible consequences of this effect be further investigated. As long ago as 1975, scientists in Stockholm demanded the retention of tritium in nuclear facilities. Today, there is still no economic way of doing so.[112] Tritium can be incorporated into virtually any molecule which normally contains hydrogen, and is also contained in the cell fluid. If it gets into thymidine, one of the molecular "building blocks" which make up the chromosomes, tritium contaminates the genetic material 50 to 50,000 times more strongly than if it were to have been absorbed into water.[36] According to an IAEA report, tritium in thymidine is fully 100 times more dangerous than in water, in terms of its biological damage.[83] Both the genetic effect of low-level tritium-based radiation, as well as its carcinogenic effect and increased susceptibility to infection, have been proven in mice.[18, 83]

In photosynthesis, tritium is preferentially incorporated into organic molecules. Potatoes have been designated a critical foot item. Of all foods examined, potatoes have the greatest transfer factor (i.e., uptake from the soil into the plant) for tritium, and constitute an important staple in the West.[51] Thus, tritium content in organic material is ten times higher for cows fed on contaminated grass than for those that receive it in their drinking water.[50] The Karlsruhe Nuclear Research Center found a ninefold increase in the tritium concentration of dry mass of pine needles, as opposed to their water content.[67]

Due to tritium's chemical transformation, a specific trait of radiation damage by its compounds has developed. When tritium decomposes spontaneously, it emits a beta ray and changes into helium, an inert gas. The compound of which the tritium was originally a part then becomes biologically ineffective or, depending on its structure, even toxic.[25] The genetic effect of these so-called transmutations has been proven.[30] In 1984, the *Bulletin of Atomic Scientists* published a "tritium warning."[11] Mewissen had found a new type of genetically-transmitted intestinal cancer produced by tritium in mice. This cancer occurs only in the 25th generation, and is then passed on genetically. Never before had a heritable cancer been generated in animals by means of radiation.

Ominously, due to the A-bomb tests of 1963, the mean tritium content in the rainwater in central Europe is greater than the normal level by a factor of 700 (7,000 pCi/liter, vs. 10 pCi/liter normally), according to a Swiss government response to parliamentary inquiry.[75] According to UNSCEAR 1964, peak values of 10,000 such units were found in Canadian rainwater that year, as opposed to between one and ten normally. As recently as 1982, a mean increase by a factor of 12 was observed in central Europe—since nuclear power plants, and especially reprocessing facilities are constantly releasing tritium.

The fact that very high accumulations also have to be considered in this case, is shown by the values measured in 1976 in the Karlsruhe reprocessing facility. In the tissue water of leaves from beaches, up to 29,000 pCi/l. of tritium were found, up to 455,000 pCi/l. in the air humidity, 48,000 pCi/l. In the precipitation, 576, 000 pCi/l. in the ground water, 45,200 pCi/l. in the tissue water of the spruces, etc.[138] Prior to nuclear power, the tritium content of rain was, as indicated above, a maximum of 10 pCi/l. In 1976, wine from the area around the Neckarwetheim nuclear power plant north of Stuttgart, Germany, had twice the tritium content of rainwater.[138]

In 1981, the IAEA stated that tritium in the environment could be cause for major worry by the end of the century.[44] The tritium problem cannot be wished away by statements like that of the radiobiologist, Dr. W. Burkart, at the Swiss Federal Reactor Institute in Würenlingen, who assures us that tritium can be ignored since it contributes only a few millionths of a rem to an individual's radioactive exposure.[15]

It is misleading to trivialize these releases by averaging, since locally high exposures may be "smoothed out of existence." It is precisely these local high concentrations to which living things are usually more sensitive. Thus in 1979, Weiss et al. demonstrated that increasing numbers of local tritium and krypton sources from nuclear power plants can be discerned from Western to Central Europe. Releases of up to 40 times the local background radiation from A-bomb fallout were achieved.[137] Pine needles show significantly-increased tritium contents, up to ten times higher, for distances as far as 10 miles from nuclear facilities, depending on wind direction.[119] Reichelt has pointed all this out—including the fact that model calculations of power-plant operators, based on rapid, equal dilution, are evidently no longer applicable.[85, 86]

Krypton 85

Krypton 85 is a radioactive inert gas which occurs in nature in only the most minimal quantities, since it is produced in the course of nuclear fission. It has a relatively long half-life of 10.7 years, and is a beta-emitter. Because its radiation penetrates less than 2 mm into the skin, it has been greatly underestimated. However, krypton 85 also affects the lungs, and dissolves in various ways in body fluids.[6, 136]

Reduction of its concentration in the air by dry sedimentation or wet precipitation is only minimally successful. This is a very serious factor, for 97% of the krypton remains in the atmosphere and is eliminated only by individual/independent radioactive decomposition, which may take decades.[111]

Most of the krypton 85 in nuclear power plants is retained in the fuel elements, except when cracks form. When the used fuel rods are dissolved in boiling nitric acid in a processing center to recover uranium fuel or separate out the plutonium, all the krypton 85 is released into the air. As a result, these facilities are even "dirtier" than nuclear power plants. Paradoxically, this process is what passes for "waste disposal." As early as 1975, nuclear scientists demanded retention.[112] The long half-life of krypton means irresistible build-up in the environment. The gas is heavier than air, but is eventually spread through the atmosphere worldwide. Today, atmospheric concentration of krypton 85 is millions of times higher than before the atomic age.[61]

If the krypton reaches even one percent of the maximum permissible concentration in the air (300 nCi/m^3), measurable global changes in the electric conditions of the atmosphere will begin to occur, according to W. L. Boeck, chairman of the Krypton 85 Working Group of the International Commission on Atmospheric Electricity (this was known in 1975).[6] For instance, the electrical conductivity immediately over the oceans would be increased by 43%. The electrical resistance between the surface of the earth or the water, and the ionosphere, would be reduced. It might be possible for lightning bolts in widely separated regions to be connected by electrical feedback. This may cause unexpected changes in the weather.[6]

Close to the surface of the ocean, cosmic radiation is low, in contrast to the upper atmosphere, where the air-mass absorbs most of the rays coming in from space. Natural background radiation from radon and its daughter products is also minimal. Radon is emitted only by rocks on land, and is washed out of the atmosphere by rain and snow, so that sea-air contains almost none of it. If a continental air mass moves over the sea, it is cut off from the source of radon—the rocks on land.[6]

In his articles, Seelig points out that changed atmospheric-electrical conditions, even of a short-term nature, can today be influencing tree damage.[91, 93, 94] His ideas will have to be incorporated into the research; currently, there is still very little available in this regard.[41] It was pointed out in the Cattenom Report issued by the French nationwide power company EdF, that the electrostatic effect of a nuclear power plant currently reaches "only" the order of magnitude of a thunderstorm front.[92, 93]

Forest Decline Maps and Radioactivity

Introduction

Scientific investigation by Professor Günther Reichelt has uncovered evidence to support Professor Schütt's recommendation for the incorporation of the artificial radioactivity factor into research on forest death. Dr. Reichelt has studied biology, chemistry and geography. After working as an ecologist at a government research institute for highland agriculture, he served as department chair and adviser in biology at a government teaching institute. Author/coauthor of several college texts on biogeography and ecology, he is member and an elected member of several scientific societies and member of the state advisory commission on Environmental Protection to the government of the German state of Baden-Wurttemberg, and of the state advisory commission on Protection of Nature. Reichelt made a name for himself in France and Germany with a completely novel method for detecting and quantifying forest decline—the so-called spot-check method.[78, 79, 80] French officialdom has always pointed with pride to its country's healthy forests. In response to an inquiry, the French secretary of state for forestry told the German newspaper *Welt am Sonntag* on June 5, 1983, that "The French forest is doing brilliantly." France, he added, was the country with the largest number of nuclear power plants. Only "a few acres of forest"—a dozen or so small parcels of woodland near industrial areas—were damaged.[79] On the contrary, Reichelt found that by comparison with the areas of damage in southern Germany, the levels of damage in most French forests were close to the maximum. High damage was found in the western and central areas of France, and in Brittany, not only for spruces and pines; even hardwoods like oaks and beeches showed widespread damage. Elm death reached shocking proportions, and gorse heaths were brown over wide areas.

Considering this situation, it is surprising that France has much lower emissions of sulphur and NOx than does Germany. By 1982, the figures, in millions of tons were:

	sulphur	NOx's
France	2.4	1.3
West Germany	3.5	3.0[110]

Moreover, France has twice as large an area, a much lower industrial density, and the west winds from the Atlantic Ocean carry little pollution.

Reichelt points out that the observed damage is most likely explained by a synergistic effect with radioactive emissions. Attempts at understanding the actions of photo-oxidants only, particularly ozone, are insufficient, he says. Reichelt believes that the conditions for the production of ozone, especially in the areas of heavy damage near the coast, are not sufficiently explained by ordinary "photosmog." He therefore urges that "the task of uncovering the basic mechanism producing the damage be addressed immediately, since this is the heart of the whole problem of forest death."[87]

Reichelt's Novel Spot-Check Method

This method is superior to the usual forest-damage surveys which are based on area averages according to forest blocks or grids. In the first Sanasilva Study in 1983 (the damage survey for the Swiss forests), Switzerland was divided equally into squares, and the overall damage was ascertained regardless of tree species (see grid on page 144).

This grid method has a number of disadvantages. If, for instance, a square contained no evergreens, but only deciduous trees, that square may appear overly healthy; for evergreens at the same location, damage would probably have been discovered as early as 1983.

Reichelt operates according to a principle analogous to weather-maps, in which all locations with the same temperature or air pressure are connected by lines. He connects all points in a forest with the same level of damage by lines he calls "isopleths." The selection of locations is not arbitrarily pre-established by means of a grid on a map; rather, the spot-check areas follow the natural shape of the areas containing the most common species of tree, such as spruce. For every sufficiently large stand suitable for a spot-check, a "mean damage level" is calculated from a sample of 10-20 graded trees. After entry onto a map of enough spot-checks, lines of the same "mean damage level," or isopleths, can be drawn.

The grid on page 145 shows a novel isopleth charting of the French region of Alsace, from the Vosges mountain range to parts of southern Germany.[79] The darker the areas, the greater the forest decline. In the spring of 1984, the Strasbourg Forestry Office confirmed the damage ascertained by Reichelt; until then, everything had been denied.[69] In 1985, after 80% of the forest area had been damaged, the communities in the Vosges Mountains appealed to Paris for government aid.[144]

A comparison of the two types of maps shows that sources of damage (emitters) can be pinpointed much better by using Reichelt's method. Charting on the basis of a fixed grid, on the other hand, does not provide a clear pattern of damage, partially because of the limited size of the forest areas. As in the case of meteorological maps, it is not necessary for forest-damage surveys to designate an observation point ever quarter

Development of forest decline in 1983. Surveys in the summer of 1983 and the Sanasilva Immediate Program as of autumn 1983, regarding firs, spruces and beechwoods.

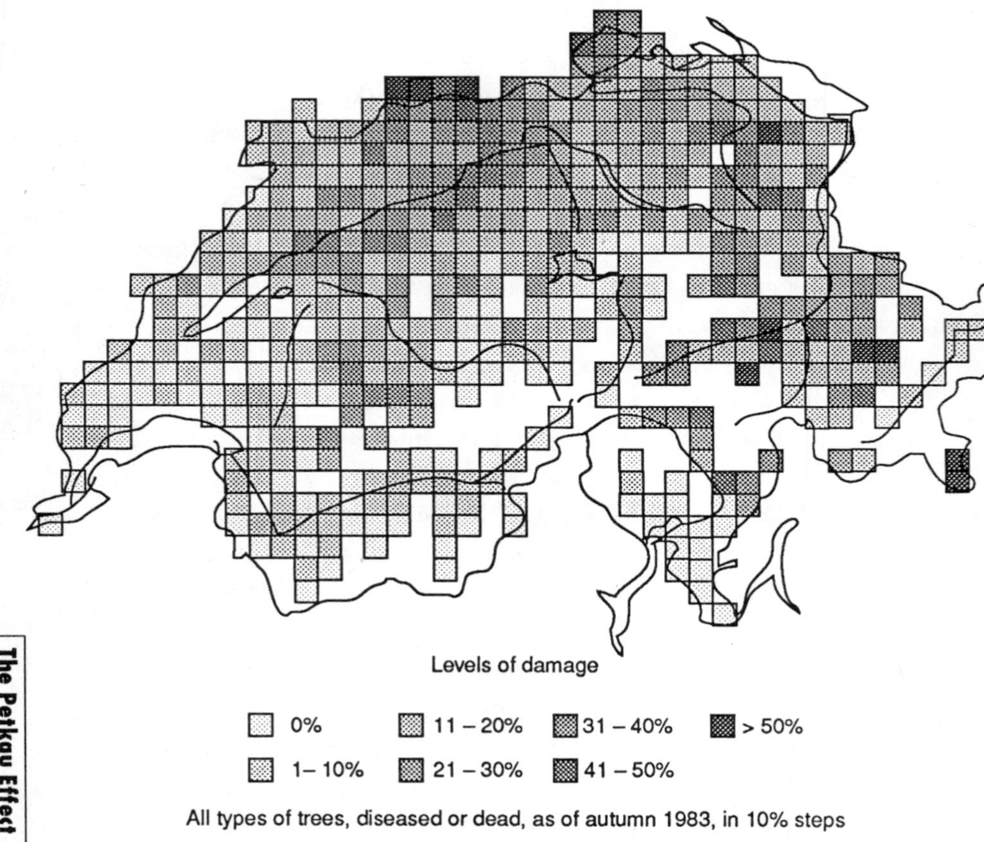

All types of trees, diseased or dead, as of autumn 1983, in 10% steps

mile or so. The pattern of damage can be detected more readily in their regularity and individuality by means of the isopleths.[82]

Reichelt's method has by now been accepted by government agencies and experts in the field.[78, 79, 86, 145] The new Swiss Sanasilva studies are being carried out by the spot-check method.

Reichelt's Forest-Damage Profiles for Classic Fossil Fuel Plants

As long as Reichelt and his charting led to conclusions regarding fossil pollution sources only, everything was fine.[78, 79, 80] No one objected. By means of

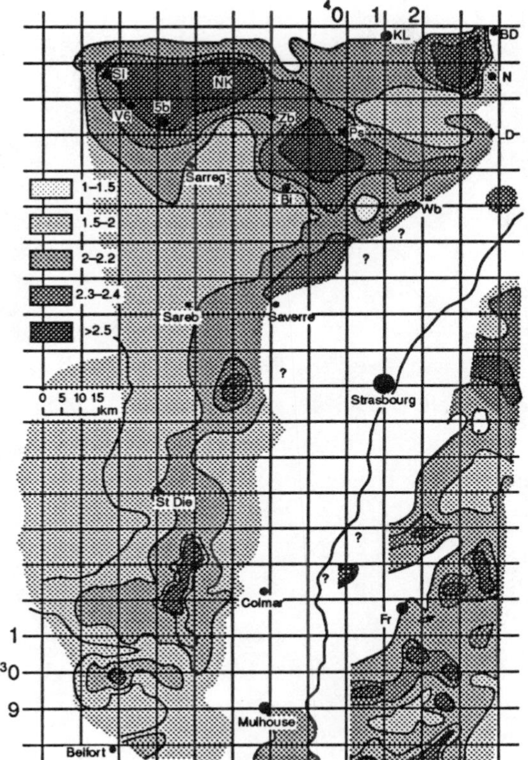

patterns in forest decline or damage isopleths, known industrial plants, which emit SO_2, NOx, hydrocarbons, fluorine and chlorine compounds (particularly refineries, power plants, chemical plants, brick-kilns and porcelain plants), could be readily identified. Reichelt's charting indicates that measures to reduce emissions from fossil sources (SO_2, NOx's) are necessary, and that many forests can be saved if such measures are implemented in time.[85]

Regarding his survey in France, Reichelt mentions that the coeffect of photo-oxidants such as ozone could have a major effect not only locally, but throughout the central mountain region. What is surprising, however, is the high damage level in western and central France, the origins of which are still unknown.[79]

The magazine *Natur* reported how impressed many scientists in the field were with Reichelt's work.[68] Professor Mohr of the Biological Institute of the University of Freiburg, Germany, called it a "scientific study of high quality." Professor Havlik of the Meteorological Institute at the Aachen Technical College said it was "absolutely virgin territory," and Professor Weischet, also of Freiburg, noted that it was the "first important step in uncovering the cause" of forest death.

Reichelt's Detection of Forest Decline near Nuclear Plants

In recent investigations, Reichelt found spots of damage near nuclear power plants in France and Germany. These damage patterns could not be explained by the pollutants. The earliest indications of damage came primarily from studies around the German nuclear power plants at Obrigheim, Esenshamm and Wurgassen, and the French heavy-water reactor at Brennilis in Britanny, and the Bugey nuclear power plant near Lyons. It should be emphasized: these were not chartings targeted at the nuclear facilities.[21, 35, 68]

The map below shows conditions for spruces ascertained at the Obrigheim nuclear power plant in Baden-Wurttemberg south of Frankfurt).

The darker the surface coloring, the greater the forest damage. The length of the axes of the wind compass correspond to the frequencies of winds at the Buchen station, 25 kilometers from Obrigheim.

During a targeted charting in May 1984, Reichelt found that a similar damage pattern could be ascertained around the Würgassen nuclear power plant in Westphalia, south of Hannover, as reproduced below.[83, 88]

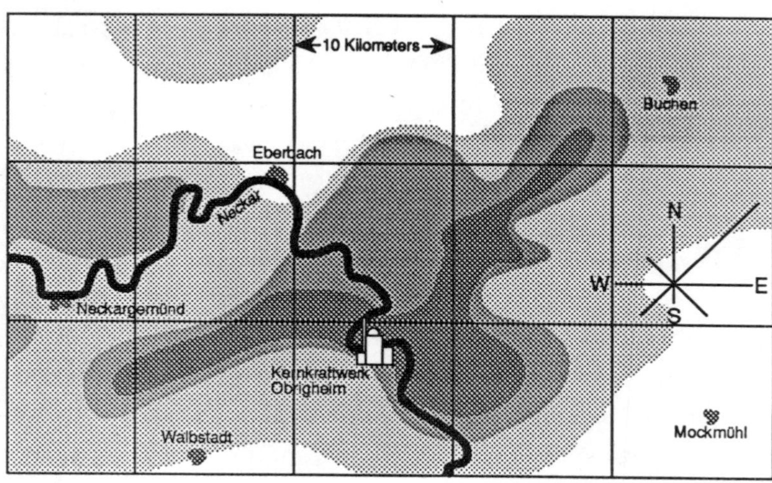

Reichelt later presented the hypothesis that the Würgassen nuclear power plant was very probably the main causal agent. A number of statistically-supported arguments would in fact, prove his case beyond any reasonable doubt.[88]

The map below shows Reichelt's results. Once again, the darker the surface coloring, the greater the damage to the forest.

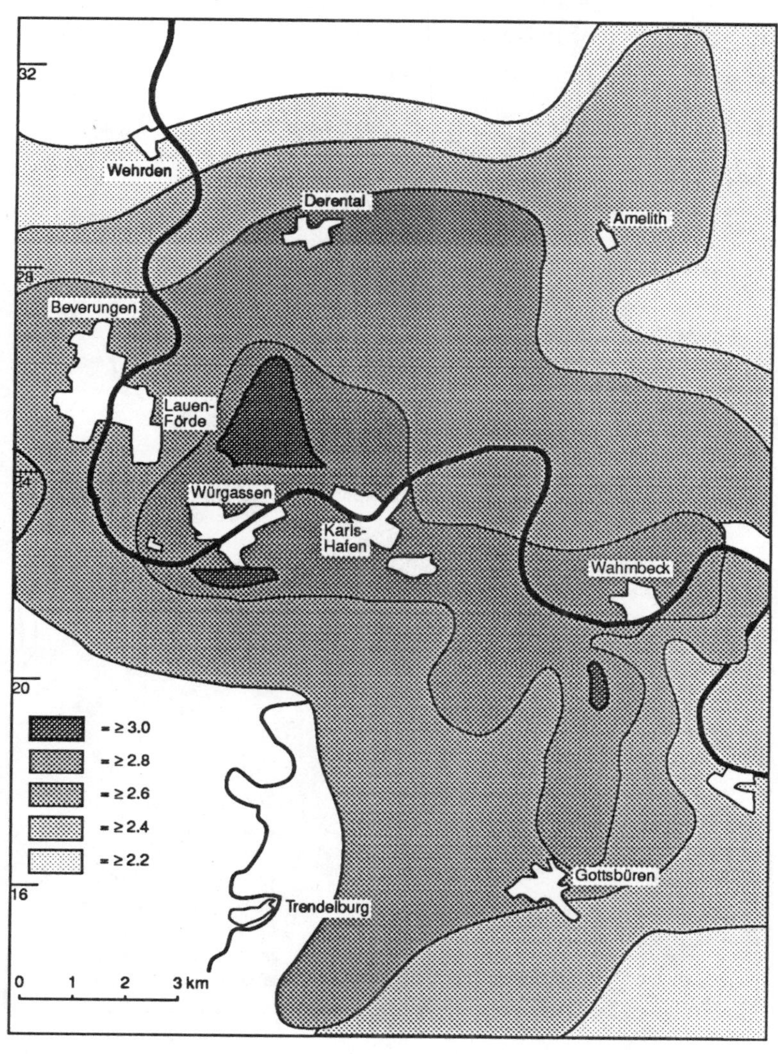

Forest Damage Profiles of the World Wildlife Fund-Switzerland

In the late 1980s, the World Wildlife Fund had become aware of Reichelt's chartings. The Forestry and Environmental Planning Office in Rudolfstetten, Switzerland, undertook a damage survey of the forests around the Muhleberg, Gosgen and Beznau nuclear power plants. Two university forestry engineers developed damage maps according to Reichelt's method. Once again, peak values in the immediate vicinity of nuclear power plants, and extensive damage plumes were found. The reports stated that the spatial correspondence of nuclear power plants and forest decline is so close that "a causal connection must be assumed.[142]

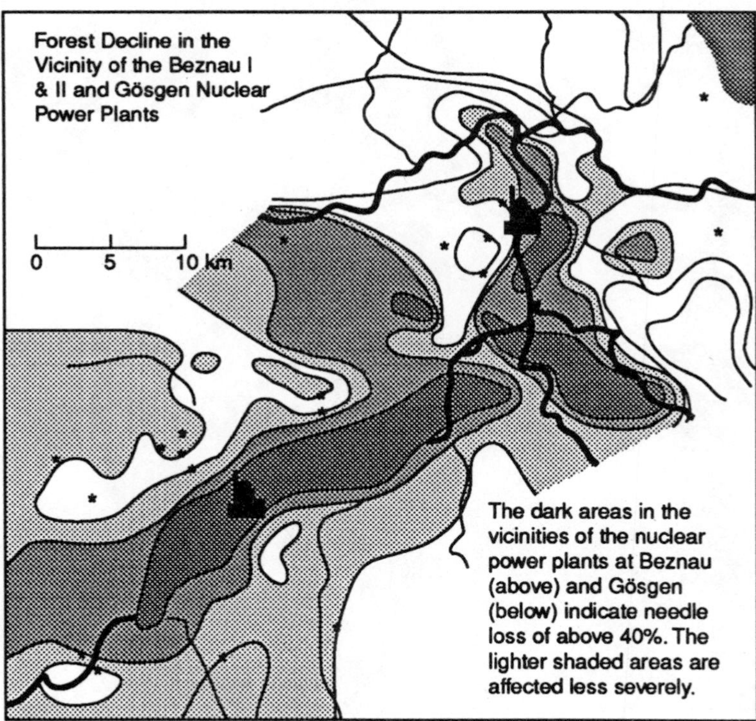

The dark areas in the vicinities of the nuclear power plants at Beznau (above) and Gösgen (below) indicate needle loss of above 40%. The lighter shaded areas are affected less severely.

Forest Decline Due to Nuclear Technology?

The publication of the damage profile of the Obrigheim nuclear power plant in *Natur* and the *Basler Zeitung* caused an uproar among nuclear advocates.[35, 67] They reacted without proper preparation. Reichelt was slandered in the usual manner. The Obrigheim plant charged that Reichelt had ascertained forest decline at locations where there weren't even any forests (Barth).[1] In fact, spot-checks had all taken in forests, after which isopleths had been drawn between points with the same level of damage. The map shows overall geographical damage patterns, in this case for spruce, regardless of whether there actually is spruce at a particular location.[81] The critics did not even have the requisite technical knowledge to understand the nature of Reichelt's technique.

Reichelt was also accused of giving the wrong wind directions for Obrigheim.[1] The magazine *Natur* erroneously stated that the wind indicator was applicable to Obrigheim. In his original manuscript, however, Reichelt had expressly pointed out that he was referring to Buchen, 25 km away. At that time he had had no access to an Obrigheim wind compass, which is slightly different. Thus, according to a report the nuclear power plant uses, the wind in Obrigheim blows mainly toward the east in summer, but toward the west in winter.[69]

However, wind direction at the emission site is not the only determinant of the distribution of pollutants. Weather-inversion patterns and dispersion conditions involving weak, or even no wind, may play a part in the formation of the damage pattern.[98] In winter, the resistance of trees to damage is less effective.[138] Later, publication by the Swiss Federal Environmental Protection Office in Bern stated that on June 26, 1984, Reichelt had successfully refuted Barth's charges at a press conference in Bern.[1, 145]

Using theoretical considerations, the Karlsruhe Nuclear Research Center also attempted to refute Reichelt's findings, in two studies entitled "Nuclear Technology and Forest Death"[53] and "Environmental Radioactivity and Nuclear Technology—Possible Causes of Forest Death?"[54] Commenting on these theoretical models, Reichelt says:

> For instance, they generally exclude errors of the measurement method at the emission and immission sites, they take only incomplete account of dynamic processes—especially when nu-

merous factors are involved—and they can therefore at best constitute approximations that do not have the same force of proof or weight in decision-making when compared with actual observed effects. . . . In any case, deviation of the actual damage pattern from theoretically predicted behavior do not make the damage pattern less credible, but should rather lead to a reexamination of the model calculations and their presuppositions.[87] Reality does not necessarily fit the model calculations.

Beyond this, the Karlsruhe objection is based on a study of various deciduous and Evergreen trees which were irradiated externally by high gamma doses—again, at a high dose rate.[100] Thus, 50% of oaks die when irradiated with 3,650 rads over the course of 16 hours; the figure (the so-called LD50 dose); for pines it was 692 rads. (The LD50 value is the dose required to be lethal for 50% of the irradiated population.) The report reassuringly states that the damage declines at lower dose rates. This and other theoretical calculations are then used to support a conclusion regarding the harmlessness of low doses and dose rates, i.e., of the impossibility of forest decline due to radioactive emissions from nuclear facilities.[53, 54]

The study cited by the Karlsruhe group in fact states that much information is still needed in order to make reliable statements on the effects of radioactive fallout. In addition, insufficient knowledge of beta rays and their possible interactive effects with gamma rays are emphasized. This makes the extrapolation of the effects observed at high dose levels to low dose rates in the case of fallout, even more problematic. Obviously, forest death does not involve lethal external doses, nor the question of whether fewer trees will die from lower external doses. According to the stress hypothesis, which is our point of departure, much more than radiation is involved in the discussion—including that for trees, only the general resistance to disease and other stress need to be affected. Even delayed radiation damage cannot be excluded. In fact, the Karlsruhe study notes that radiological knowledge is insufficient for making predictions regarding harmful effects on growth and the yield of future harvests. The LD50 dose provides too little information on this and the entire range of radiation's harmful effects on humans and animals.[100]

Moreover, the Karlsruhe report refers to a 1981 IAEA study to investigate the behavior of tritium in ecosystems (e.g., transfer factors, incorporation into organic material, biological half-lives), and to determine the biological significance of thee radioisotopes under various natural conditions.[53, 54] The surface exposures were high in this case too—three to five orders of magnitude above those permitted at nuclear facilities. According to Karlsruhe, no damage to plants was observed. Therefore, forest decline due to the tritium exposure from nuclear operations need not be considered. The Petkau Effect, of course, is not mentioned. Nevertheless, the study expresses the apprehension "that tritium could, around the turn of the century, give cause for great worry."[44] An entire catalogue of crucial research is presented to illustrate how meager our knowledge is, and how great our fears, regarding the tritium which has been released so thoughtlessly into the environment. Possible synergistic effects—the complex interactions of pollutants multiplying their effects—were not taken into account.

The response of Dr. Wolfgang Ehmke, a Green Party member of the German Bundestag, to a parliamentary inquiry in the summer of 1984 makes clear that such investigations are considered undesirable. The response stated: "The Federal Government sees no further need for research on the correlation between forest death and nuclear power."[21]

The fact that low doses can be more harmful than high ones had already been discovered in 1967, with respect to the possible effect on growth rates of hela cells in water containing tritium, and the growth of roots in vicia faba.[37] At the high dose rate of 32 rads per hour (0.32 Gy) there was only half as great a damaging effect as at the low dose rate of 0.5-3 rads per hour (0.005 0.03 Gy). Nonetheless, the later is 10,000 times higher than the dose rate of background radiation. It should therefore be expected that as a result of the Petkau Effect, minimal dose rates will be more harmful in plants than would be expected. The lack of significance of such experiments with very high dose rates is shown by the spider wort example. The BEIR Report of 1972 reported that a spider wort, irradiated at the high dose rate of 30-40 rads per day for 15 weeks, is affected.[3, 99] Yet, the Japanese scientist Professor Ichikawa Sadao, of the Agriculture Department of the University of Kyoto, found spontaneous genetic mutations at a rate 30% above the average among spider worts planted near nuclear power plants where dose rates are thousands of times lower.[45] In this species, genetic mutations are particularly easy to spot by changes from blue to pink in the mature pollen filaments.

Moreover, significantly increased mutation rates in spider-worts were found as far as 10 km southeast of Esenshamm, on the north coast of Germany nearby the Lower Weser Nuclear Power Plant.[63, 85] At this plant, Reichelt also found indications of damage to stands of spruce.[85]

Thus, even in plants, there is concrete evidence that experimental findings at higher dose rates provide misleading information. This must be investigated with great urgency.

Reichelt's paper concerning Obrigheim was accepted in January 1984 by the *Forstwirtschaftlisches Centralblatt,* a highly-respected forestry journal. The article appeared in September of that year in spite of criticism from the nuclear advocate experts.[1, 53, 61, 85] This should indicate the invalidity of criticisms to Reichelt's findings.

Forest Decline around Mines Containing Uranium

One of Reichelt's most important studies of radioactivity and forest death reports on mines containing uranium. Reichelt mapped the vicinity of the Wittichen mine in the Black Forest in southwestern Germany, as well as six other mines containing uranium in the Fichtel Mountains of the Czech-German border, the nearby Upper Palatinian Forest in Bavaria, and the Black Forest. His results show that damage is greatest in the immediate area of the mine, and decreases beyond a distance of about 2.5 km.[69, 84, 86, 87]

The figure on the following page shows the mapping of the old silver and cobalt mines near Wittichen, which contain uranium. The darker the surface, the greater the damage to trees.[84, 86]

In spite of relatively minor SO_2 and NOx contamination of the air, young spruces are yellowed and crippled. However, the air contains increased radioactivity due to the naturally radioactive inert gas, radon, which is emitted by uranium.

Areas around Wittichen have particularly high damage. Here, the damage falls off in all directions—not concentrically, but according to irregular patterns. On the map it appears as if a major emitter were located in the little hamlet of Wittichen. Actually, the village's few houses consist of single-family homes and a small convent; no corporation emitting pollu-

Forest Decline Map around the Mines with Uranium Content at Wittichen in the Black Forest

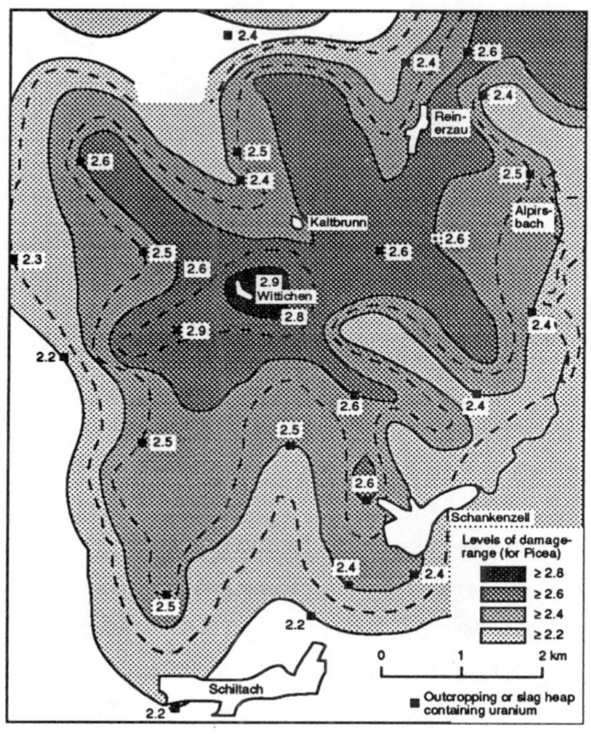

tants, no factories. The findings were confirmed by the evaluation of infrared images from an aerial survey conducted by the Forestry Experimentation and Research Institute in July 1983, 1984 and 1987.

Reichelt made similar observations in mines containing uranium, both in France and in northeastern Bavaria. "In some cases, forest-decline observations actually constituted successful detective work, since unexpectedly high forest damage led to the discovery of mines not marked on the maps."[82]

Reichelt believes that the effect discovered in 1975 by Vohra, Director of the Indian Nuclear Center in Bombay, is operative here. Namely, radon causes a rapid transformation of sulphur dioxide into sulphur trioxide—hence, into sulphuric acid.[86, 88] Even a low radon content of 50 picocuries per cubic meter of air is enough to trigger a chemical reaction.

Such concentrations occurred in Wittichen near one of the slag heaps, and may have actually exceeded this amount.[88] Here, in 1982, Reichelt was able to make use of measurements conducted by Schmitz et al. on the basis of which he calculates values of 11,000-135,000 pCi/M^3.[105, 106]

According to Vohra, the minimal required SO_2 concentration in the air is only 5.2micrograms per cubic meter. The mean level in Wittichen, 20 micrograms per cubic meter, also exceeds the minimum.

Reichelt assumes that the toxicity of SO_2 becomes much more acute in the presence of radioactivity. It may be increased dramatically when the radioactivity acts together with aerosols and fog, or with rainfall. The prior dry deposition of pollutants on needles and twigs will have an additional amplifying effect. In this regard, Reichelt refers to Subba Ramu et al., 1981, who demonstrated that reactions sparked by radioactivity progress much more drastically if ozone and hydrocarbons are added in moist air.[88, 102]

It should be noted that radioactivity by itself cannot be considered the sole cause of forest decline.[88] Photos taken in 1956 show that Wittichen once had particularly beautiful fir and spruce stands. Even earlier, slag heaps were, for the most part, covered with trees. Since 1962, however, large-scale forest decline has been occurring. Evidently, the damage is caused by the interaction of radioactivity from mines, and the minimal quantities of fossil-fuel pollutants, such as SO_2, in the air.

Government Research Contract

Significantly, on August 3, 1984, Reichelt received a research contract from the Ministry of Food, Agriculture, Environment, and Forests of the state of Baden-Wurttemberg. He was funded to conduct mappings of forest decline in selected areas of Germany.

Unfortunately, an application by Reichelt to perform a much more comprehensive study was rejected; he had wanted to conduct a study together with the noted plant physiologists Professor H. Metzner of the University of Tubingen, and Professor F. Fezer of the Geographical Institute of the University of Heidelberg. Parallel with the mapping, aerial images using infrared radiation detectors were to be made, and leaf and soil samples were to be systematically analyzed for radioisotopes and other chemical pollutants.[88] Was there a fear of this carefully targeted, compact and comprehensive research project?[1] As time passes, proof of radiation damage will be more difficult to ascertain due to the removal of sick trees. Metzner only received a contract for a literature survey on "radioactivity and damage to plant organisms."[66, 88]

In April 1985, Reichelt's work was completed. By then, he had had a chance to discuss his work with other scientists. The mappings had been carried out between April and October of 1984, with 760 spot-checks being done for the isopleth maps. Reichelt's results demonstrated among other things, that increased forest decline was occurring at all mapped sites (seven nuclear power plants and a uranium ore slag heap), even where other significant emissions could be excluded.[88] He also found indications that emissions from nuclear installations and industrial sites reinforced one another's effect on forest damage (nuclear power plants at Stade on the north coast, Grundremmingen in Bavaria and Beznau, Switzerland).[88] It is interesting to note, as does Reichelt, that wherever government aerial photographs of the areas he had gone over were available, his damage-profiles are confirmed.

Reichelt has recommended continued research, but has stated that in order to ensure its independence, it should not be conducted at nuclear research centers.

Radioecological Observations of Forests

Almost Everything Is Still Unresearched

It is always assumed in laboratory experiments that damage in plants which grow up in natural ecosystems, always depends on numerous factors, though given the same radiation dose exposure.[100] It is not possible to work with radiation exposure alone.

Thus, from an ecological point of view, one decisive factor is the quantitative changes relative to the original situation, that is, prior to the impact of civilization. Of all civilization-induced pollutants, radionuclides have had the highest rate of increase relative to their natural rate of production.[138]

The military and civilian use of nuclear energy is responsible for the massive increase in the level of radioactive pollutants in the environment. Contaminants include radionuclides (radioactive pollutants), which previously existed in nature only in the most minimal quantities, or not at all.

For instance, the atmospheric concentration of krypton 85 has increased by millions. In the case of tritium, the increase in Switzerland was 700 in 1963, and still 12 in 1984 (annual mean values). By contrast, the classic fossil pollutants have increased only by factors of two or three; while regional increases by factors up to nine have been found in the case of NOx—in Switzerland, for example, from 1950 to 1982.[23, 82]

Such annual mean values can be deceptive, and are certainly not authoritative in and of themselves. Living things react to peak exposures during critical periods, not only to annual-mean-value exposures; the latter can indeed be "averaged" out to indicate no danger at all. It would be like placing one foot into ice-water, the other into water heated to 80°C (175°F), and claim that no harm will occur because the person is standing in water of a comfortable "mean temperature" of 40°C (105°F).

In a recent literature review conducted by Metzner for the Ministry of the Environment of the state of Baden-Wurttemberg, 800 titles were evaluated. (Metzner, director of the Institute for Chemical Plant Physiology of the University of Tübingen, is recognized as one of the leading German experts in the area of the use of radioactive substances in biochemical research on the basis of numerous studies.) Metzner established that radioisotopes of the inert gases can selectively accumulate in particularly sensitive areas of the plant, such as in cell membranes—the carriers of the genetic information of a cell, and in such other cell organelles as chloroplasts and mitochondria.[147, 66] This is similar to iodine's concentration in the thyroid. If the possibilities of selective accumulation are ignored, the resulting damage to the organism will be impossible to predict. Finally, the accumulation of radioactive substances in needles and leaves could, due to the falling of dead leaves, lead to a contamination of surface areas of the soil. This would cause damage to the very sensitive root hairs. The radioactivity could then return to the plant through the roots, leading to a vicious cycle of damage.

Metzner suggests additional research in order to investigate the radioactive accumulation. This research contract has since been awarded to Professor Münch of the University of Heidelberg.

The concentration mechanisms of artificial radioactivity are absolutely crucial. For instance, large accumulations of radionuclides and their daughter products in leaves, needles and the soil due to the bomb fallout of the '50s and '60s have been proven with certainty. This is also true for plutonium, americium, tritium and carbon 14. These accumulations in some cases amount to increases of three to five orders of magnitude (factors of 1000 to 100,000) over the levels of natural sources."[82, 39] These figures would be inconceivable for ordinary chemical pollutants.

Dr. Armin Weiss, director of the Inorganic Chemistry Institute of the University of Munich, and a member of the Bavarian State Legislature, has now determined that forest death has been strongly increasing throughout the Northern Hemisphere since the time that the long-lived radionuclides from the A-bomb tests started spreading to the root areas of trees.[87, 138] Reichelt, too, points out that many nuclides that were incorporated first into the leaves and forest trash have now become available to the roots. Thus, the observed delay of the damage might well be connected with this phenomenon.[82, 87]

The Petkau Effect in Plants

There is every reason to fear that the Petkau Effect, discovered only in 1972, manifests itself in plants as well as in animals. As in the case of countless radiobiological animal experiments, research had been done only with radiation at high doses and high dose rates—i.e., in the entirely wrong ranges for exposure to low-level environmental radioactivity.

The 1972 BEIR report states:[3]

> There is little or no data on the effect of chronic radiation on plants. Lilium and tradescentia were damaged with 30-40 rads per day (duration of irradiation: 15 weeks). . . Evergreens such as pines and taxus were influenced at approx. 2 rads per day.[99] Chronic effects were found in oaks irradiated for 10 years with some 7 rads per day.[64]

Whicker, too, reports on radiation experiments in the wild with far too high doses levels.[140] In forests, a reduction of biomass production and narrower growth rings were observed for gamma radiation down to the level of one rad/day. Abnormal growth, reduced germination and reduced seed growth have also been reported. However, in some cases, increased seed production was found. Here, however, reduced competition with other plants, or changed environmental conditions, played a role. For instance, dying trees can improve the competitive situation of more radiation-resistant trees and plants by increasing their sunlight exposure.

In the same vein, Zavitovski reported on a large-scale study by the AEC, designed to ascertain the effect of ionizing radiation on typical North American forest ecology systems.[143] The plan was to irradiate plants over the course of five sequential summers. A gamma source consisting of 10 kCi of cesium 137 was installed. The trees were irradiated for 20 hours daily, from 30 minutes past midnight to 8:30 AM. Within the test range, at a distance of up to 150 meters, the total radiation dose was between 60 and 50,000 rads. The daily dose was approximately 0.5 to 500 rads. In a remote control area, "non-irradiated" trees were likewise observed.

Rather than conducting the irradiation for the originally-planned five summers, it was conducted only during the summer of 1972 (May through October); in 1973, the AEC called off further irradiation, supposedly for reasons of cost and the establishment of new priorities. However, the forest was observed carefully both prior to the irradiation (1972) and thereafter (1973-4).

The effects of radiation were demonstrated in dead or dying trees, and in premature or delayed sprouting, leaf-coloring, leaf-falling, blooming and fruit-bearing. The most common characteristic is overly long vegetation-latency, thinning of leaves are also reported. Some of the effects are only reported one or two years after irradiation, or took that long to develop fully. Was it for this reason that the observations were broken off early—for fear that reports would now start to yield interesting results on delayed damage? It should be added that various species of trees show very different susceptibility to radiation.

Though a frighteningly large number of damaging effects showing parallels to current forest death are described in the study, it is based on dose rates that are far too high, and periods of irradiation that are much too short. Possible damage resulting from years of chronic radiation damage at low dose rates can not be determined. In addition synergistic effects of radiation and fossil pollutants were ignored altogether.

In contrast to animals and humans, no experiments have been conducted on plants to determine the consequences of the Petkau Effect. Although there is discussion of the combined damaging effect of SO_2 and NOx and also of ozone, on the cell membranes of trees, the effect of radioactivity is excluded from the debate. Yet, the Petkau Effect may play a key role in analyzing/decoding the complexity of the recent types of damage to the forests.

It should also be remembered that natural radiation is based on alpha-emission, and is in terms of the Petkau Effect, less important than artificial beta-emitters such as tritium. The concentrated ionization effect of an alpha beam on the cell plasma will enable many more of the dangerous oxygen radicals that are generated to recombine and thus, become ineffective. This is in contrast to the case of a low-energy beta beam, which permits more radicals to avoid recombination and thus, have a greater chance of reaching cell membrane and of damaging it.

A study by the Brookhaven National Laboratory in New York also refers to the very insufficient radiobiological data available for plants. Data is particularly lacking in the evaluation of damage from beta radiation, and of possible synergistic effects on the plant kingdom involving gamma radiation.[100]

Sensational Government Studies

In a Swiss Federal Interior Department report of September 1984 on "Forest Death and Air Pollution," the insufficient degree of radioecological knowledge is admitted for the first time:

> Since, on the other hand, most investigations into the biological effects of radiation have concentrated on humans, not all questions regarding the effects on entire ecosystems such as the forest can yet be answered.[23]

Schütt had previously pointed out that Reichelt's research had initiated a series of experiments—albeit in the face of opposition from the nuclear research establishment—that would officially clarify a connection between radioactivity and forest death.

A great sensation occurred in July of 1985, when a publication by a government environmental protection agency in Bern confirmed, for the first time in the world, Reichelt's findings.[145] This amounted to a disavowal of all nuclear advocates who had previously criticized Reichelt, and who, without exception, had claimed with such great certainty that there could be no connection between forest decline and nuclear technology. One critic was, of course, the then-West German federal government.[21]

This publication by the Swiss Federal Office for Environmental Protection in Bern not only explicitly recognized Reichelt's method on damage evaluation, and his cartographic representation by means of isopleths, but also determined the following:

> In the vicinity of a number of nuclear power plants and facilities (mines), greater damage may appear than in comparable areas without nuclear facilities. The level of damage is comparable with that of industrial emissions.[145]

This statement was confirmed by investigations implemented by the World Wildlife Fund in the vicinity of Swiss nuclear facilities.[142, 145] The study does, however, contain a proviso informing that the general validity of the statement is still open to question, and draws the following conclusion: "The studies evaluated do not as yet permit any conclusive evaluation of the question at issue. Too many questions are still unresolved; too many hypotheses are on the table." The belief is expressed that with targeted research work, "the broad outlines of whether and to what extent radioactive emissions from nuclear power plants share responsibility for forest decline, could become clear."[145]

Still, nuclear advocates in government and business are not going to simply throw in the towel. We are starting to see the same game that was played with the infant-mortality statistics and the cancer mortality statistics in the U.S.A. Now that Reichelt's data can no longer be denied, a new tack will have to be tried. The Forestry Experimental and Research Institute in Freiburg Germany, has undertaken computer calculations in order to "correct" for certain special factors in Reichelt's damage data. For example, the exposed positions of the damaged forests or particularly disease-prone stands of trees in Obrigheim and Wittichen might be revised. Not unexpectedly, the revisionists reached the conclusion that no extraordinary damage exists, but refused to make the calculations public.[147]

General Perspectives

In current scientific literature on forest death, all classic pollutants and their effects, as well as their distribution in the environment, are discussed. Limits of tolerance are being sought for emissions, usually for individual components. This presupposes that the effect can actually be predicted, and that equal pollution of the air will result in equal damage, independent of the presence of other factors. This "causality principle" is what is used for logical evaluations in day-to-day life.[9] Causality is the concatenation of cause and effect. If we know the existing condition of a system and all the effects to which it will be subjected, it will be possible to predict its future. A physical system which repeatedly begins a process under the exact same conditions will always behave the same—equal causes will have equal effects.

Modern physics, with its new "chaos theory," has demonstrated that this "principle of causality" and its way of viewing things is not often valid.[19] Many systems are extremely dependent on initial conditions, and even minimal deviations will cause unsystematic, "chaotic" results. Similar causes no longer have similar effects. For example, weather predictions cannot provide any great certainty, since conditions in the weather-forming atmosphere represent a chaotic system.[19] All computer calculations are helpless in this situation.

Dr. J. B. Bucher of the Forestry Institute in Birmensdorf, Switzerland, writes in the forestry journal *Forstwissenschaftlishes Zentralblatt* (4/1984) that the phenomenon of emission-caused forest death can become plausible on the basis of the chaos theory.[9] "Even minimal differences in natural environmental parameters (conditions) such as weather or variable conditions due to location will have decisive and unpredictable effects on the emission effects on plants, given similar levels of air pollution."[9]

Today, it is generally assumed that natural pollutants can be safely increased up to a certain threshold. Though this may be true for certain components, given the presence of even two pollutants, it may no longer hold true. Thus, when trees are fogged with NO_2, the effect is "fertilizing" (i.e., positive), but when this occurs in the presence of SO_2, the effect is growth-impeding, or negative.[9]

According to Bucher, as a result of this chaos-principle it may be impossible in certain cases to provide a scientific causal explanation of forest death, in which nothing remains except epidemiological proof.[9] (In the case of forest death, this would involve forest decline profiles using such methods as mapping or aerial photography.)

It is now being demanded that the environmental pollution levels of classic pollutants be reduced to that of the 1950s. This should certainly be attempted as rapidly as possible using appropriate filter facilities, catalytic converters, and regulations regarding the properties of the fossil fuels. At issue is our very lives.

Yet even this may be insufficient. According to computer calculations of the Institute for Applied Systems Analysis in Laxenburg, Austria, forest death may actually be accelerating. With more and more damaged trees, and fewer and fewer trees overall, less pollutants are being filtered out of the air. Even with reduced emissions, the pollutant content of the air would remain the same, or even increase.

However, the Institute's report unscientifically exclude the effects of radioactivity from the outset. Nuclear technology, which carries a share of the blame, cannot simply be "forgotten" in forest decline research. The first signs of forest death coincide with the first appearance of artificial radioactivity in our environment. Therefore, our highest priority must be to seek knowledge on the effect of radioactivity on our environment.

The Soil is Dying

In recent years, the protection of the soil has been recognized as a vitally necessary objective. Major symposia have been held, reports have been written that our soil-protection concepts have been insufficient.[148] Dr. O. J. Furrer of the Swiss Federal Agricultural Chemistry and Environmental Health Research Institute, says:

> The soil is a fairly stable ecosystem. . . . For this reason, therefore, damage due to pollutants rarely appear there in the short term. That, however, makes the long-term effects all the more ominous, for pollutants continually accumulate in the soil and eventually lead to acute damage. Soil thus contaminated can in many instances no longer be restored; the damage is irreparable,

the soil is lost, dead. In establishing pollution limits for the soil, a level of tolerance for the total content of the soil has until now been the point of departure, with the proviso that this content level would not be attained in the soil over a long period, such as 100 years. Usually, too, what was taken into account was not total pollution from all sources; rather limits for only certain sources, such as air, sewage, water, etc., were established.[31]

Furrer then demands that the soil be protected not for a certain period, but *forever*; also, various soils have differing capacities for the immobilization of pollutants, so that all pollution sources have to be taken into account—rain, air, commercial fertilizer, pesticides, sewage, garbage compost and dust. Long-term protection is possible, Furrer adds, only if the material cycle is in balance.

Bucher believes that current forest-deterioration suggests that there can be no such thing as a pollution-threshold—the so-called "no-threshold principle," implying a zero-emissions goal. The consequences do not inevitably lead to chaos. Scientific models of societal transformation based on a new ethic can be worked out. This will mean an environmental policy in which preventive measures are oriented toward maximum-risk assumptions.[9] In a modern society, nuclear power would, of course, not be tolerated. It is obvious that we have to ban the release of artificial radioactivity. It cannot be biologically broken down in the soil and, depending on complex factors, it can continue to accumulate in the biosphere for extremely long periods of time.[3]

IV. Fundamental Sociopolitical Consequences

Dr. F. A. Tschumi, professor of environmental biology at the University of Bern, announced many years ago that three important pillars of our society are based on foundations that are no longer environmentally sound: the ethics of the individual, the educational system and the economic system.[125, 126]

The Questionable Ethics of the Individual

The modern human being recognizes a responsibility primarily toward the individual, the family and the state—i.e., the second level of the ecological ladder. By contrast, his or her responsibility toward humanity, and toward the ecosystems and the biosphere, is badly lacking. Prior to an understanding of ecology, it was believed that "life" was manifested in individuals.

However, the individual is dependent on ecosystems. For this reason, family planning has become a precondition for the protection of life; without it, no environmental protection measures will have any useful effect. Family planning should operate in accordance with the ethics of responsible parenthood, and above all, this ethic should take effect before new life has begun. Thus, there is a superordinated responsibility to protect the ecosphere, and ultimately the biosphere. We will have to face up to this necessary "planetary responsibility." This will mean accepting the wonderful cycle of nature, and working to reincorporate ourselves into it.

In spite of great intelligence and multifaceted knowledge, most people are much too ignorant of their new responsibility. Moreover, there is not yet sufficient social cohesion. Neither the political, scientific nor technological communities recognize any obligations.

Misdirected Education

Our misdirected education plays a major part in the deplorable condition of our environment. Our education system neglects to inform us on the natural foundations of our civilization. Much too little instruction is concerned with environmental factors.

Too little attention is paid to ecological interrelationships which are indispensable for an understanding of the current environmental situation and for formulating a future-oriented mode of thinking. Ecology should be a mandatory subject in all schools.

Every young citizen, no matter what profession he or she chooses, gains access at school to more information and skills than ever before. Yet, these youths are not being equipped with an understanding of how to deal with our environmental problems.

Recently in political circles, there has been much talk of seeking a balance between the economy and ecology. For the most part, this involves nothing more than well-meaning lip-service, since the preconditions for truly effective measures are not in place. There are still too many obstacles thrown up by sociopolitical power hunger and the legally-entrenched status quo. Most scientists are trained to become the willing partners of business and the technocracy. They think far too little about whether their work upholds responsibility toward biological life and toward the future. Certainly, reevaluation is beginning, but so far, it has hardly had any practical effect. These scientists mistake their frog-in-the-well outlook for a valid world-view. Specialists are needed whose vision is not narrowed by specialized knowledge and dependency relations, and who are capable of providing the basis of expertise for interdisciplinary thinking within the framework of ecological consciousness. A new concept of education is needed. Nuclear technology would not have a chance in such a framework.

Faulty Economic System

The final problem is a fundamentally false economic order. It ignores the vital ecological principles mentioned earlier: it tries to do without a self-contained, closed cycle. By ignoring renewable energy, it denies both producers and consumers a decomposer for the recycling of their wastes. On this basis, only a small portion of humanity can afford to live in ever greater opulence. The statement "the rich get richer and the poor get poorer" cannot be dismissed as a political slogan; it is a sad reality. By enriching ourselves we are endangering forests, diversity of plant and animal species, productive soil and a clean atmosphere, on a global scale.

Today, we dream up scenarios in which people hardly have to do any more manual work. Take, for instance, the following passage written in 1972 by Eugene Rabinowitch, former editor-in-chief of *The Bulletin of Atomic Scientists*: "The only animals whose disappearance might threaten the viability of humans on earth are the bacteria which normally inhabit our bodies. As for the others, no convincing evidence exists that humans could not survive as the only animal species on earth. If economic procedures could be developed for the synthesis of food from inorganic materials—and this will be possible sooner or later—humans may even become independent of the plants on which they are currently dependent for food." Evidently, Rabinowitch never heard of ecology.

Such misguided, purely technological-economic thinking—the product of misdirected education—will have to be reoriented. Technology and civilization are only meaningful when they are not directed against nature, but rather achieve a state of harmonic interaction, which tolerates the limits of our ecosystems and our own health.

The possible disappearance of forests, animals, and plant species cannot simply be expressed in monetary values—losses to the economy. By simply calculating the costs of forest death to the economy of mountain ranges—and it would run into the billions—we would still not grasp the essential problem. The diversity of species and the basis of all life is at stake.

The existing nation-state-based economy will soon be confronting its complete ruin. In 1972, Dr. C. Binswanger, professor at St. Gallen College in Switzerland, described the situation succinctly. In his view, the interdependence of ecology and economy implies new dimensions in economic theory. Binswanger points out that even Aristotle had distinguished between two types of economy—economics based on nature, and economics opposed to nature. While the former was an economy oriented toward the supply of food and other vital necessities, the economics opposed to nature was comparable to a commercial money economy. Binswanger felt that natural economics should once again have great significance in world commerce. This would involve, first of all, an incorporation of the ecosphere. Economy and ecology would merge to form a unity which can only be vaguely imagined today.[5]

Professor K. W. Kapp of the University of Basel believes also that "national economy has failed to anticipate the effects of modern technology on the environment." For this reason, the economy faces "radical new tasks." Economic systems should be seen as open systems requiring a respect for ecological questions, and hence, interdisciplinary thinking.[5]

Far-sighted economists have long understood that a new consciousness expressed in ecological terms is necessary. They recognize that "life," including human life, does not depend on technology, but rather on the proper functioning of natural ecosystems.

This does not mean that economic systems and industry should simply be destroyed—quite the contrary. Rather, the limits of technology and value-free science have to be made public. Only an educated population will be willing to accept the many sacrifices it will have to make. Moreover, it is not merely a question of self-denial, but of a conscious turning away from artificially-generated needs. No one will want to pay for a higher standard of living with sickness, and lingering death.

In understandable language, citizens must be presented with the results of science from the point of view of holistic ecological thought. There should be much more discussion of the wonderful biological cycles and the incomprehensible intelligence of nature. There should be much less focus on supposed economic necessities, and of classic environmental pollution and poisoning.

The Chernobyl disaster at the end of May 1986 should make it clear to everyone that the risk of nuclear power is unacceptable. This disaster has clearly shown the insidious nature of radioactive fission products,—as well as our own powerlessness to protect ourselves against it. The unacceptability of nuclear power should be disseminated in the mass media and in schools. A planned phase-out of nuclear power should be demanded. The foundations of our health, which are rooted in biologically active soil, must also be discussed.

Chernobyl and a thorough education of the population will logically result in drastic economic and social consequences. An educated public will support these consequences, for they will guide us into a better future—and away from catastrophic destruction. Most urgent is the adoption of a new responsibility within the framework of an ecological protection of nature. This book has attempted to make such a contribution.

V. POSTSCRIPT

Increased Crime Rates Produced by Radioactive Fallout?

The research of Sternglass and Bell on mental retardation, discussed earlier, has prompted psychologist Dr. Robert Pellegrini, Professor of Psychology at San Jose State University, to investigate the criminal behavior of persons born in the 1950s and 1960s—the period of atmospheric atom bomb testing. Persons between the ages of 15 and 24 exhibit the highest rate of criminal violence in the United States today.

Pellegrini suspected that the growth in criminal offenses in the total population would be at its maximum 15 to 24 years after the period of heaviest fallout. Afterwards, a drop should occur. Pellegrini found that this was precisely the case. In the following graph, the percent change in the criminality rate per 100,000 persons is compared for each decade with the previous one from 1956 to 1985.

As Fig. 1 shows, in 1956–1965 relatively few individuals in the critical age of maximum criminal activity between 15 and 24 years were exposed to fallout early in their growth, since they were born between 1932 and 1951. Those who were exposed to fallout in the 1950s had not yet reached the critical age. From 1956–65, aggravated assault increased only 23%, whereas forcible rape and criminal homicide actually decreased by about 9% each.

The next decade, from 1966 to 1975, contains the greatest number of individuals exposed during early development and who had reached the critical age of 15 to 24. All three types of violent crime increased in a highly significant manner in comparison with the previous decade (1956–1965).

From 1976–1985, the number of 15 to 24 year-old individuals irradiated during infancy declined once again with the end of atmospheric

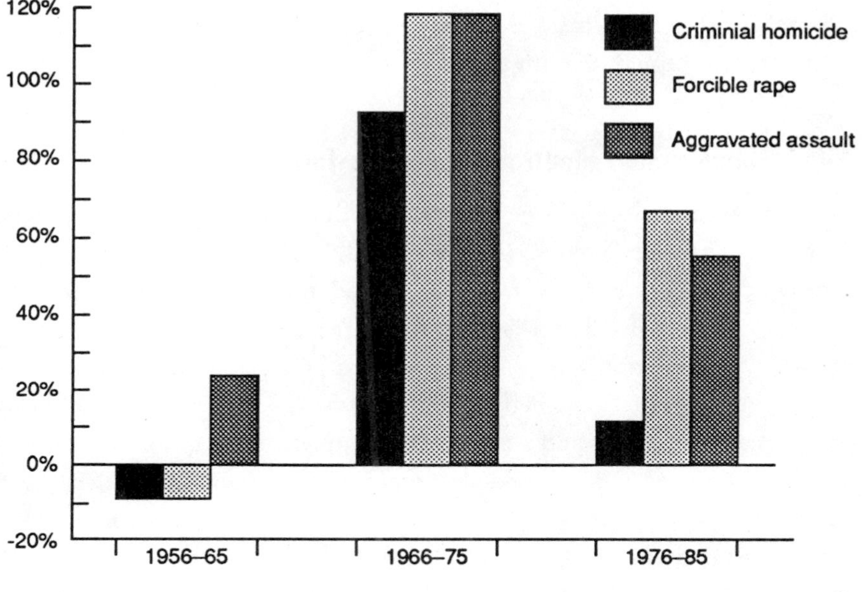

tests by the U.S.A. and U.S.S.R. in 1963, and, in comparison with the previous decade (1966–1975), the violent crime rate dropped markedly. Nevertheless, there remains an increase in comparison with the period 1956–1965. Pellegrini is of the opinion that individuals exposed to radiation in early childhood could remain predisposed to some criminality during their whole lifetime, and not only during the critical ages of 15–24 years.

Pellegrini does not maintain that radioactivity or radiation is the only cause of criminal behavior. On the contrary, he refers to the research of Hollinger, Öffner and Ostrov (1987) that looks for a connection with sociological, psychological and pathological factors. With his work, Pellegrini introduced a fourth and certainly startling factor that will have to be taken into account in future research.[1,2] Indeed, as has already been demonstrated in animal experiments, exposure to radiation in early development can result in more aggressive and retarded offspring.[3]

Mental retardation and also low IQs were particularly prevalent among descendents of atom bomb victims in Japan whose mothers were

8–15 weeks pregnant at the time of the blasts.[4] According to the ICRP (Thorne) and BEIR V, the dose response relationship for these effects appears to pass through the origin in a linear manner—i.e., no threshold dose exists.[5, 6]

AIDS Epidemic and Fallout: A Possible Co-factor?

Various causal factors for the outbreak of the AIDS epidemic have been proposed. The most likely hypothesis is still that the AIDS virus originated through the mutation of a similar virus found in green monkeys of tropical central Africa from where it spread to other parts of the world. This hypothesis is supported by a number of facts. For example, a Norwegian family was found to have had an HIV infection going back to 1965, based on the medical records and stored blood samples.[10] The father, who was a sailor, had visited various African ports in the preceding years. Furthermore, antibodies could already be identified from central African blood samples as early as the 1950s. Finally, the virus could have existed for at least 40 years, according to studies by a prominent research laboratory, the Dana-Farber Cancer Institute in Boston. [7] Forty years ago no gene technology existed. Therefore, the assumption that the AIDS organism could have arisen from gene manipulation in a laboratory and subsequently got out of control is untenable.

Until now, AIDS research has not come up with an argument to disprove the hypothesis that radioactive fallout plays a role in the development and spread of AIDS. This hypothesis was developed in 1986 by Professor Ernest J. Sternglass at the University of Pittsburgh and Professor Jens Scheer at the University of Bremen.[12,13] Both researchers point to previously inexplicable aspects of the AIDS epidemic: the sharp increase in AIDS cases in 1980-82 and the initial concentration of cases in central Africa, the Caribbean, and the east coast of the U.S.A.

According to Sternglass and Scheer, during the atmospheric bomb tests in the 1950s and early 1960s, the irradiation of bone marrow by beta rays emitted by strontium 90 and other bone-seeking fission products could have caused a harmless, pre-existing HI-like virus to mutate into a deadly form—the present HI-virus.

Stokke and co-workers discovered in animal experiments that the doses of strontium 90, as low as 10-100 mrem, bring about a significant reduction in the number of bone marrow cells. [14] With fewer bone marrow cells, the bone marrow produces fewer blood cells for the immune defenses or results in mutated cells unable to perform their normal function. This form of low dose-rate damage can be explained by the Petkau Effect.

During the period of massive atmospheric bomb tests, beta-rays produced by strontium 90 affected a large number of individuals born around this time. Their developing immune system was weakened before or shortly after birth as a result of the irradiation of the bone marrow "stem-cells" from which all forms of blood cells develop. Furthermore, strontium 90 causes important blood cells which are produced in the bone marrow (the so-called NK or natural killer cells) to become ineffective.[15]

Sternglass and Scheer believe that the acquired immune deficiency caused by strontium 90 during intrauterine development is the most dangerous long-term effect of fallout (see also Dr. R. Jensen, "Umweltschaden AIDS"). [16]

In 1980–82, some 18 years after the peak of the atmospheric bomb tests, a sharp increase in the number of AIDS cases occurred. Young people carrying the latent mutant virus had become sexually mature. Consequently, individuals born with an impaired immune system were exposed to all forms of sexually transmitted diseases especially rampant in the cities of central Africa. This brought about the multiplication of the T-helper cells in the blood, which are vital in the immune defense system, and in turn gave rise to the rapid multiplication of the HI-virus.

In other words, the virus uses the T-helper cells for its own reproduction. When the T-helper cells die, the viruses escape into the blood stream. Thus, under favorable conditions, the AIDS virus can spread quickly through the blood, for example by sexual contact or directly, by the use of infected needles.[17] Such conditions prevail precisely in the poorest regions of the world like central Africa.

Ninety percent of fallout, including strontium 90, comes to the earth as a result of precipitation. The AIDS epidemic started in areas with high rainfall such as Central Africa and the Carribean—near the geographical latitude of the bomb test sites in the Pacific. It appeared first and at the greatest rate in the high rainfall regions of the west and east coasts of the United States. Far fewer AIDS cases have been reported in the dry region of northern and southern Africa and inland U.S.A.

Though subject to higher rainfall, southeast Asia showed relatively few AIDS cases. Rice and oceanfish contain less strontium 90 per gram of calcium than do milk, bread, meat, potatoes, vegetables and freshwater fish, which are the dominant foodstuffs in North and South America, the Caribbean and Africa. A United Nations report (UNSCEAR) for 1962 showed that out of 22 countries investigated in 1957, the highest levels of strontium 90 in bone were found in young individuals in tropical central Africa.

In March/April 1988, the German newspaper *Wochenend* dispatched a team of journalists together with a doctor who specialized in tropical medicine to Zaire.[18] According to the report, there was widespread incidence of extramarital sex, rituals with monkey blood, abundant monkey meat (from the green long-tailed monkeys) offered for sale, and hardly any instruction about safe sexual practices. One in every three persons in some areas of central Africa is believed to be carrying the virus. The thesis concerning the origin of AIDS in tropical Africa is therefore strongly supported. According to the team of journalists, official reports would be expected to either falsify or hide the true state of affairs. Professor Sternglass has called attention to the fact that , as early as 1958, Nobel Laureate Andrei Sakharov had expressed the fear that viruses could be mutated into dangerous forms as a result of fallout.[19] Sakharov had also predicted the crucial role of free radicals in weakening the human immune system.[20]

Revision of Dose Levels Are Totally Insufficient

After some ten years of resistance, the ICRP finally commented on the revision of the radiation dose calculated for the Japanese atom bomb victims. Excessive rates of cancer have been appearing among these people for as long as 40 years after exposure to the bomb radiation. And the linearity (or supralinearity) of the dose response is now known to hold for doses down to at least 0.2 Sv (20 rem).

Todays scientists debate whether curve a or b is valid in the region below 0.2. Sv (see figure), both c and d being no longer consistent with the mounting data for low dose effects. The re-evaluation of the dosimetry has revealed that low doses of radiation are likely to be six to ten times more dangerous than previously considered by the ICRP.[21] As

Figure 2
Various Dose Response Relations

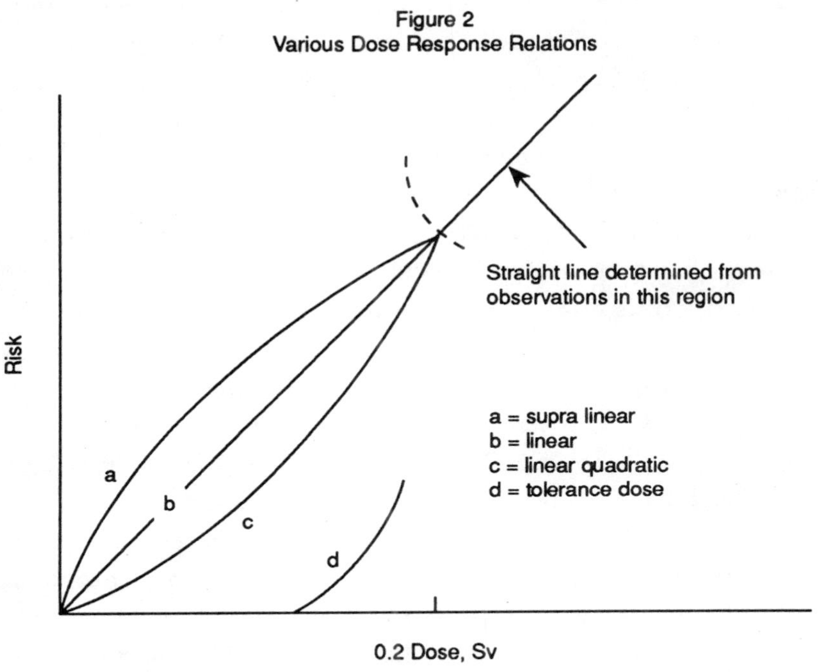

recently as 1977, the risk was calculated to be only 100-125 cancer deaths per million persons from a dose of 10 mSv (one rem).

In the fall of 1987, the Radiation Effects Research Foundation (RERF) reanalysed the Hiroshima/Nagasaki cancer statistics. It reported a new risk factor of 1,300 deaths per million with a linear model.[22] Professor John W. Gofman found a risk factor as great as 3,000 deaths from cancer per million exposed to 10 mSv.[22] At the same time, Professor E. P. Radford (a past chairman of the BEIR committee) declared that the cancer risk from radiation incidence was at least ten times greater than that which the ICRP had previously adopted.[22,23] Despite these findings, the ICRP postponed a decision to revise its risk estimates at its congress in Como in 1987. The British National Radiation Protection Board disagreed, and immediately recommended a reduction of the maximum permissible dose for radiation workers from 50 mSv (5 rem) to 15 mSv (1. 5 rem).

Karl Z. Morgan, past president of the ICRP for many years, himself took the ICRP to task in 1986. [24] He complained "that the ICRP action delayed the prevention of very large radiation hazards, that the radiation hazards were underestimated, and that the recommended maximum permissible levels were far too high. They are standing afresh at a crossroads: from a linear dose response curve to a supralinear, downward moving curve which clearly shows a much higher risk at low doses." As explained above, such a relationship is also produced by the Petkau Effect.

In November 1990, the ICRP finally issued its new recommendations.[25,26] The basic maximum permissible level of 50 mSv (5 rem) per year for radiation workers was reduced to 20 mSv (2 rem), i.e. a factor of 2.5. The value of 20 mSv is nevertheless not a rigid limit—it is to be the average over five years. In any single year, according to the ICRP, the maximum radiation level can remain at 50 mSv (5 rem), as before. This "averaging," according to Serge Prêtre, Swiss delegate to the IRCP, is itself a "sweetener for the atomic industry." Prêtre believes that the nuclear power plants can cope with this limit.[25] However, this means more automated robots, more staff, more time for repairs, as well as higher costs. In Switzerland this new radiation protection regulation will first come into force in 1993. Yet to be discussed is whether people who are exposed to radiation under 30 years of age should have a lower limit set of 15 or 20 mSv, since recent studies clearly show that young persons are more sensitive.[26] This is apparently what Prêtre hopes to see adopted in Switzerland.

These new, lower levels are nothing but a cosmetic action. A tenfold reduction at the very least would be necessary. In a recent American-German study, Professor Wolfgang Köhnlein of the Radiation Biology Institute of the University of Münster and Dr. Rudi Nussbaum, Professor of Physics at Portland State University, demolished the scientific basis for assuming that relatively slight radiation effects occur at low doses. For the average risks of the Japanese survivors in the dose range from 0-490 mSv (0-49 rem), an additional 1,750 plus-or-minus 350 cancer deaths would be expected if one million persons are exposed to one rem. This is a study which is based on low doses, thus rendering an extrapolation from high to low unnecessary. The study also shows a supralinear dose-response curve.[27,28,29,30]

The above dose reductions and the calculations on which they are based (Hiroshima-Nagasaki studies) are, from the very outset, unsatisfactory. This was recently pointed out by the Otto Hug Radiation Institute in Bonn in a 1989 report.[22] Five noted professors from German universities are signatories to this publication. (The five are Prof. Inge Schmitz-Feuerhake, Dept. of Medical Physics, University of Bremen; Prof. W. Köhnlein, Radiation Biology Institute, University of Münster; Prof. H. Kuni, Dept. of Nuclear Medicine, University of Marburg; Prof. E. Lengfelder, Radiation Biology Institute, University of Münich; and Prof. R. Scholz, Dept. of Biochemistry, University of Münich.)

They noted that the observations of increased numbers of cancer and leukemia cases after the A-bomb tests in fallout areas such as southern Utah and in the vicinity of nuclear plant installations such as Hanford, Rocky Flats and Sellafield cannot be explained by using the Hiroshima risk coefficient.[22] The professors also corroborated many of the results obtained by independent researchers, which this book has reported.

In February 1990, the Medical Research Council of Britain published a report in the *British Medical Journal* in which a significantly higher incidence of leukemia in children living in the vicinity of the Sellafield reprocessing plant was observed. In the neighboring village of Seascale the cases were up to ten times higher than the national average.[31] Even the U.S. Department of Energy (DOE) has changed its previous stance: independent qualified scientists can now obtain access to data from 600,000 persons who have worked since the early 1940s in atomic weapons production—for many of whom, increased risk of cancer had already been reported in smaller studies. As discussed earlier, Robert Alvarez had already requested this information in 1984 when the Japanese study was unable to provide the necessary data for the formulation of low-level radiation standards.

The report prepared by the Otto Hug Radiation Institute also calls into question whether a purely external radiation exposure—as experienced by the Japanese—can be compared at all with an exposure due to ingested or inhaled radioactivity. The complexity of the biological process is much too simplified by making such a comparison on the basis of equivalent doses as is done in the official radiation-protection standards.[22] In a similar manner, the recent BEIR V report now draws attention to this difficulty as discussed below. "The rem lie" that was described earlier is thus exposed.

The Otto Hug Radiation Institute report also draws attention to the variation in individual radiation sensitivity, a fact that is generally not taken into account in the radiation standards (e.g. age, sex, health, race, genetic disposition).[22]

Genetic defects, as well, are greatly underestimated. According to Dr. Peter Weish of the Austrian Academy of Sciences, Chernobyl will be the source of genetic defects on a continental scale.[32]

Recently, by using mutations induced in mammalian cells in the laboratory, it was demonstrated that an inverse dose effect appeared even in the cell nucleus as long as the dosage or dose rate was set low enough (Crompton, 1985; Kiefer, 1986).[4] This means that a small dose given at a low rate can produce more mutations than a higher dose given more rapidly (the inverse dose rate effect) (see figure).

This result argues in favor of a much higher risk—not only for membrane damage, but also for genetic defects—than would be expected on the basis of a linear extrapolation from high doses given at high dose rates.[4]

When compared to the normal dose response curve for acute

Figure 3
Dependence of induced mutations on dose rate for mammalian cells (V79). When compared the normal dose response curve for acute (short-term) gamma radiation, the curve for the higher dose rate of 50 mGy/h is flatter. If the dose rate is reduced to 8 mGy/h, the dose curv response curve rises more rapidly with increasing dose, producing a greater than normal eff

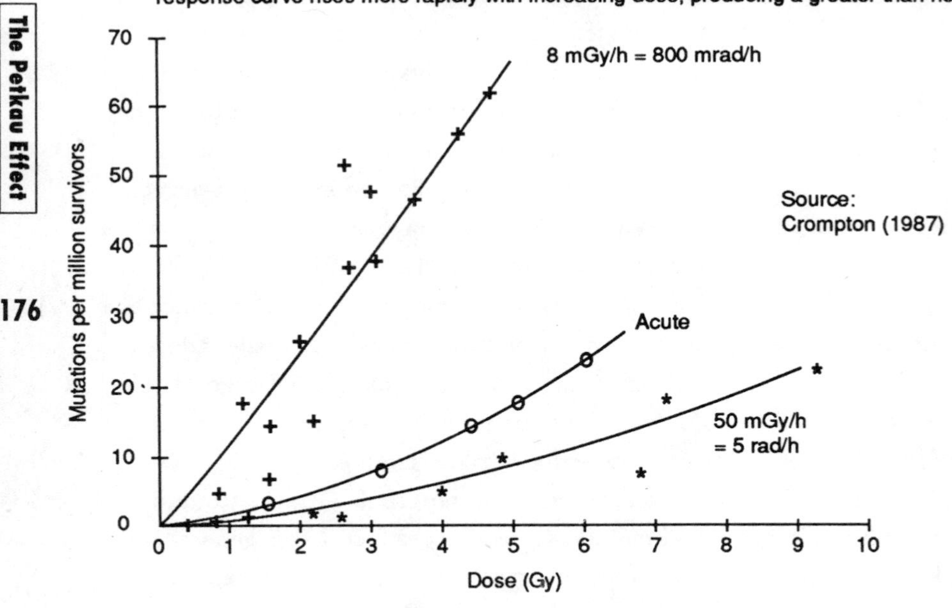

(short term) gamma radiation, the curve for the higher dose rate of 50 mGy/h is flatter. If the dose rate is reduced to 8 mGy/h, then the dose response curve rises more rapidly with increasing dose, producing an effect greater than normal.[4]

The Proceedings of the National Academy of Sciences also support these findings. A 1986 paper published by the Academy indicated that the human mutation rate at small doses is some 200 times greater than one would have expected, based on the experience with higher doses.[33] (This paper was made possible by new detection methods for genetic damage that had only recently been developed.)

The Latest Concerning the Petkau Effect

New research shows that the oxygen-free radical plays an important role in cell damage and could be the cause of many diseases.[34,35] This research relates to both the outer cell membrane as well as the cell nucleus. Of course, free radicals are produced not only as a result of radiation, but also through biochemical reactions. Chain reactions occur in the cell membrane, whereby the efficiency of peroxidation of membrane phospholipids increases as the dose rate is reduced (the Petkau Effect).[4,22] According to present knowledge, the aging process is coupled to this lipid peroxidation process. Lengfelder refers explicitly to the Petkau Effect in this connection.[4]

In 1988 I presented a lecture during a symposium at the University of Münster in Germany on the "reality of the Petkau Effect."[35] During the discussion, the scientists favoring atomic energy were unable to disprove my statements with a single argument. They alleged that Petkau had worked with artificial membranes. In fact, a large part of his work demonstrates the opposite.[36-48] He, himself, had found evidence for the Petkau Effect not only in living microorganisms and mice, but also cited evidence for the inverse dose rate effect in human lymphocytes (white blood cells) in the studies of Kenning (1984) and Müller/Rayleigh (1982) in a 1986 publication.[46] Petkau also spoke of the reality of the inverse dose rate effect in human diseases and furthermore provided evidence for a connection between cell membrane damage and the development of cancer.[42] Chernobyl is a tragedy, but also a unique large-scale experiment in which the serious biological consequences—which independent epidemiological researches in the U.S.A. and Germany make plain—can be explained only in terms of the Petkau Effect, as will be discussed in detail.

According to the BEIR V report, two-thirds of all biological damage at low dose rates in the cell nucleus results from the effects of free radicals when weakly ionizing radiation such as X-rays, gamma rays and beta rays, are involved. Likewise, in his book *Deadly Deceit: Low-Level Radiation, High-Level Cover-Up,* Dr. Jay M. Gould refers to the Petkau Effect in his book over and over again.[49]

In 1989, when the laboratory management cut off all funds for his research team, Petkau resigned his post as head of the Medical Biosphysics Branch of the Canadian Atomic Energy research laboratory in Pinawa, Manitoba. Presently an associate professor in the Department of Radiology of the University of Manitoba and the author of 92 papers in the field of radiational biology, he has opened his own medical practice.

In a letter to the author, Petkau summarized the essential results of his research work as follows:

> The inverse dose rate effect that I have described in refereed journals applies to 1) Radiation induced breakage of model phospholipid membranes by Na^{22} (*Health Physics*, (1972) [239–244]).[22] The phospholipids were extracted from fresh beef brain. The membranes were planar in structure. These types of membranes were also used to show that superoxide radicals are attracted to their surfaces (*Can. J. Chem.* 49 (1971) 1187–1195). The collection of superoxide radicals at the membrane surfaces also changed the radiation chemistry in the water phase nearby. This insight led to the next series of experiments.
>
> 2) Radiation-induced lipid peroxidation in spherically shaped model membranes made from phospholipids extracted from soybeans (Biochim biophys Acta 433 (1976) 445–456). Irradiation was done with an external Cs^{137} source. The inverse dose rate effect was observed when the membranes were unprotected by superoxide dismutase (SOD), a scavenger of superoxide radicals. However, when SOD was added, the membranes did not undergo a measurable amount of lipid peroxidation. Hence, no dose rate

responses were observed. These results clearly implicated superoxide radicals in radiation induced lipid peroxidation of phospholipid membranes. A concurrent Hungarian study on mice showed that SOD also fully protected against lipid peroxidation in the liver induced by an external source of X-rays. Thus, these in-vitro results were consistent with each other.

When tritiated water was used to irradiate these membranes, I found that the dose rate relationship could be extended down to background levels, thus demonstrating for the first time a specific radiation-induced chemical process at background dose rates in a biologically relevant system. The tritiated water was in effect an internal source of radiation. I found that the inverse dose rate effect was modified by SOD but, in contrast with the Cs^{137} study, not totally eliminated. Thus, these studies showed up a fundamental difference in response of the membrane to external and internal irradiations.

The foregoing studies were generic in nature. Other workers have extended them to radiation induced:

1. Lipid peroxidation in human lymphocytes (reviewed in the article "Protection and Repair of Irradiated Membranes")

2. Lipid peroxidation in animals

3. Mutations in Chinese hamster V79 cells (Crompton et. al., *Naturwissenschaft* 72 (1985)(439–440).

Together, these studies suggest that humans might be more sensitive to low-level radiation than hitherto realized. . .[50]

The Death of the Forests: The Nuclear Energy Connection Becomes Clearer

As a result of an official literature survey referred to earlier and the research work of Professor G. Reichelt, the Swiss Government funded a study using infrared aerial photo reconnaissance to examine the forest damage in the area surrounding the nuclear reactors at Beznau and Würenlingen (PSI). The report of the Federal Institute for Forestry Research on this study dismissed the effect of the nuclear installations on trees "with high likelihood." However, at the same time, the report declared "that scientific proof of this result is yet to be forthcoming because the radiological evidence is not conclusive."[51]

Reichelt responded immediately that both he and the World Wildlife Fund of Switzerland had observed increased forest damage at these installations.[52,53] In August 1988 he prepared a critical study of this EAFV report and proved that the focal point of forest damage lay directly in close proximity to the nuclear installation.[54,55,56] Reichelt proceeded on the assumption that the aerial photographs made in 1985 in the test zone and the damage analysis of the tree crowns photographed from the air were carried out correctly. However, he showed that the study contained methodological deficiencies, inadequate interpretation of important factors and in particular, an insufficient analysis of the numerical and cartographic material.

Consequently, the most important map—the "relative" damage index map—was analyzed in a completely unsatisfactory manner so that it led to the wrong conclusions being drawn. Reichelt corrected this faulty methodology by overlaying the map with a grid network and computing an "average" relative damage index (see figure) for each of the 963 small squares covered with trees.

Even a purely visual examination shows at once that the focal point of the damage lies near the nuclear installations (dark squares) indicated by Beznau and PSI (Paul Scherrer Institute). By counting the squares, the damage in any area can be quantitatively compared.

The figure shows that in the high emission area (25-mR-zone with 154 squares), over 70% of the squares have above average damage, about 11% show average and 18% indicate below-average damage. Also, within a circular zone of 2. 5 km radius there was clear evidence for above average tree damage. Within this zone one finds more than half of the above-average damage, although it represents only 25% of the total area under investigation.

The distribution of damage also shows a striking correlation with wind direction. In the sectors lying in the direction of the prevailing wind (shown in the figure in the northwestern and northeastern directions), the forests showed above average damage. On the basis of the more accurate analysis of the government study, according to Reichelt, the suspicion that the nuclear installations could contribute substantially to forest death was

Mean relative forest damage index for lower Aare region.
Accumulated dose for emission range up to 25mR Argon-41.

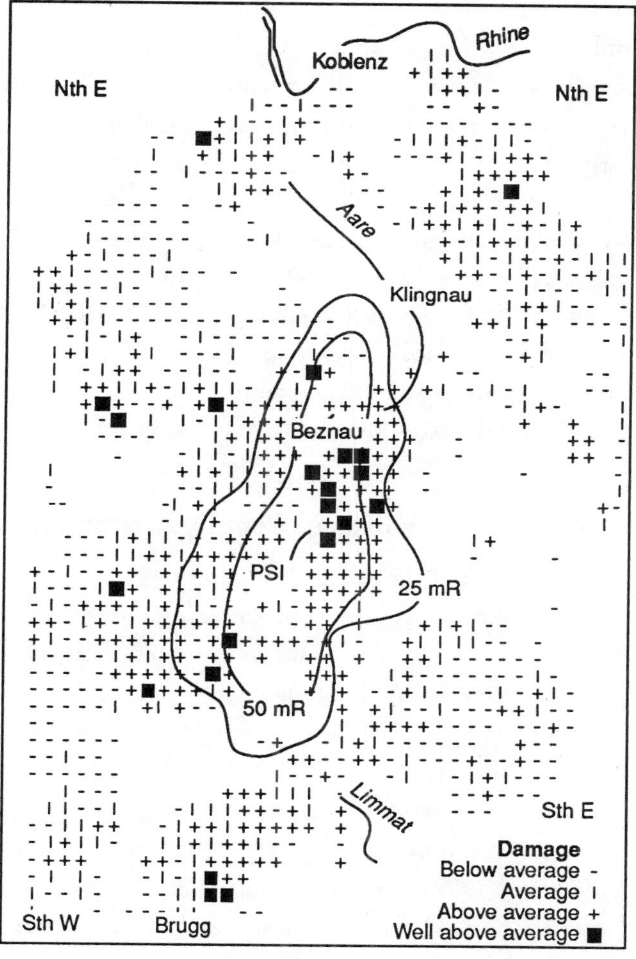

clearly strengthened. He has requested an examination of the EAFV study by an independent international authority, e.g. the European Academy for Environmental Research in Tübingen (Germany).

The EAFV study is unable to refute the three hypotheses which could explain the damaging effect of radioactivity on forests:

1. Damage to the genetic material of the cell nucleus.

2. Amplification of the effect of fossil fuel pollutants due to radioactivity (radiation smog).[52,57]

3. The Petkau Effect which first and foremost causes damage to the cell membrane.

Unfortunately, Kollert, in a 1988 publication of the University of Bremen, noted that not only German research grant applications but also those at the European Economic Community (EEC) level relating to radiation chemistry of the atmosphere (radiation smog) are being rejected.[57,58] Evidently, government agencies do not want further research into forest death related to radioactivity. Since the damage to trees has not progressed as dramatically as had been feared in 1982, some forestry scientists are going so far as to maintain that there was no dying-off of the forests after all. But this view is not shared by most forest rangers. Of course, without the image of a dying forest preying on our minds and hearts, no troublesome research in the areas of radiation smog would be regarded as necessary.

Can Insects Reveal the True Damage Done by Nuclear Releases?

Following the Chernobyl accident, Cornelia Hesse, a freelance artist specializing in biological illustrations, observed a sharp increase in the occurrence of deformed insects. She found congenital defects in insects close to atomic installations—Gösgen, Beznau, the Swiss Federal Institute for Reactor Research, Leibstadt, Sellafield, Crays-Malville, and most recently in the U.S.A. near the Three Mile Island and Peach Bottom nuclear plants on the Susquehanna River south of Harrisburg. Hesse's training allows her to make the most exact and detailed reproductions. She published her findings in 1988 and 1989 in two important daily Swiss newspapers and more recently in a book.[59,60] Her "concern with natural changes," Hesse has said, sensitized her "to ecological relationships. I noticed that leaf insects are important environmental indicators." Since then, Hesse has called for a scientific investigation into whether radioactivity could be the cause of the deformities she has observed in the bodies of insects.

Professor Willi Sauter of the Entomological Institute of the Swiss Federal Polytechnic Institute (ETH) in Zürich has also urged an investigation. According to Sauter, the "hypothesis that the defects are due to radiation as suggested by the place of discovery is reasonable." Still, it is a question which remains to be confirmed by experiments. Sauter admits "that similar damage could be caused by chemicals (e.g. herbicides). Nevertheless, the question concerning the effect of small doses of radiation is now being discussed much more than before (Petkau Effect)."[59]

As a result of such concerns, in the summer of 1989 the directors of the ETH initiated a research program examining the possibility of radiation damage to insects. Since the beginning of 1990, an ETH zoologist has been working on a dissertation in this area. Amazingly, concerned scientists—who were at first skeptical of the effects of low-level radiation—are able to undertake such an investigation, in the face of a powerful political lobby unwilling to face the facts. The industry still maintains that the doses in the vicinity of nuclear power stations are much too small to be significant and lie well below the natural radiation levels. But nature does not necessarily act in accordance with such theoretical calculations; too many have already turned out to be incorrect.

The Unsolved Problem of Hot Particles

The air masses drifting around the world from Chernobyl contained large quantities of microscopic dust particles with a high concentration of radioactivity (the so-called "hot particles"). The total radioactivity of the individual particles varied between 10-10,000 Bequerels. Their diameter ranged from 0.01-1 thousandths of a millimeter.[4] It was estimated that 5–30% of the total radioactivity was carried in the form of such small particles. As a result of the extreme heat in the fuel elements and the explosion, the reactor fuel and fission products were melted and ejected from the reactor. Some of these particles were carried to high altitudes and distributed around the globe.[4]

Hot particles have been known since the first A-bomb explosions. At that time it was alpha particles (mostly from plutonium), whereas Chernobyl contaminated the world mainly with beta rays produced by ruthenuim, zirconium, niobium, cerium, cesium and strontium. Beta rays can be incorporated more readily into organisms and result in a completely different energy distribution.

An alpha particle affects an organism only within a small distance from its source (0.05mm), and generally leads to the death of the cells through which it passes. Its total energy is absorbed by tissue in a small point-like region. Thus, as a result of the inhalation of these particles, a "hole" in the lungs can arise. This "hole", as well as the dead cells, cannot produce many cancer cells, and is therefore regarded as relatively harmless.[5] However, the theory of the harmlessness of alpha rays is very controversial.[32]

With the beta radiation from the Chernobyl particles, the dose radiation is very different. The beta particles, whose range is much greater than that of the alpha particles, pass into living cells where non-fatal doses are delivered and the cancer process can be induced.[4] Beta particles are therefore expected to be much more dangerous than alpha particles with respect to cancer and the abnormal function of cells.

Early on, the Swiss Federal Institute for Reactor Research (EIR) (Burkhart, 1987) believed that the biological effects of beta rays from hot particles were well understood. (The EIR was recently rechristened with the more innocuous name Paul Scherrer Institute [PSI].) The Radiation-Protection Authority of the Federal Republic of Germany (SSK, 1987) concurred that "hot particles do not contribute a significant fraction of the anticipated dose from the accident."[4] To this conclusion, Professor Lengfelder of the Radiation Biology Institute of the University of Münich replied that it rests on a mistaken application of the action of alpha particles.[4]

The point-like localized dose produced by an inhaled particle was calculated incorrectly, as if it were distributed evenly throughout the whole lung. In the same way, the force of a deadly blow to the back of the head cannot be regarded as distributed evenly over the whole body. Misrepresented in this way, such a blow would be predicted as harmless.

The Swiss physicist, Dr. A. Masson, has been able to demonstrate evidence of hot particles in street dust on plants, the ground, mail boxes, etc.[61] The University of Constance found hot particles in a domestic rubbish treatment plant. As a result of the re-circulation of such dust particles by various means (e.g. wind, gardening, children's play, vacuum cleaning), completely uncontrolled amounts of hot particles have been inhaled in addition to the normal dust that can be calculated by standard means. Most important, these particles are still being inhaled today.

Lengfelder points out that the risk of inhaling hot particles, particularly for those who are employed in rural areas (among other things, hay making), existed for many months and will continue to exist for a long time. Tests on roof dust in Germany showed, as late as 1988, an activity that often far exceeded 100,000 Bequerels; hot particles, in fact, continue to be detected to this day.[4]

This problem is still being extensively covered up. In the dose calculation for the Chernobyl accident, for example, the deleterious effects of hot particles have been played down. A complete investigation into the biological effects of hot particles emitting beta rays is urgently required. Such studies should under no circumstance take place at institutes advocating nuclear technology—or those getting funds from the industry.

The Fatal Summer of 1986 in the United States

The radioactive cloud which arrived in the United States in May 1986 from Chernobyl in effect turned 230 million American citizens into the unwitting subjects of a gigantic experiment, one that could never be performed on laboratory animals because of the impossibly large number required. At the same time, the U.S.A. makes public health and environmental statistics available to the public like no other country in the world. Consequently, the Chernobyl disaster can be regarded as a tragic opportunity for those interested in calculating the effects of nuclear fallout.

Dr. Jay Gould, a statistician and advisor to the Environmental Protection Agency during the Carter Administration, together with Professor Ernest Sternglass, discovered that from May to August, 1986, there was a highly significant increase in the total death rate of the population and the infant mortality rate (which includes all children in their first year of life), and a reduced birth rate.[62,63,64,49] Since all three changes were independent of one another and all three were correlated with an increase in the iodine 131 concentration in milk, the probability that these changes resulted by pure chance was far less than one in a thousand.

The next figure shows the percentage change in the number of deaths in the summer months of May-August 1986, compared with the same months of the previous year, as a function of the iodine 131 concentration in milk for the nine census regions of the U.S.A.[49, 65]

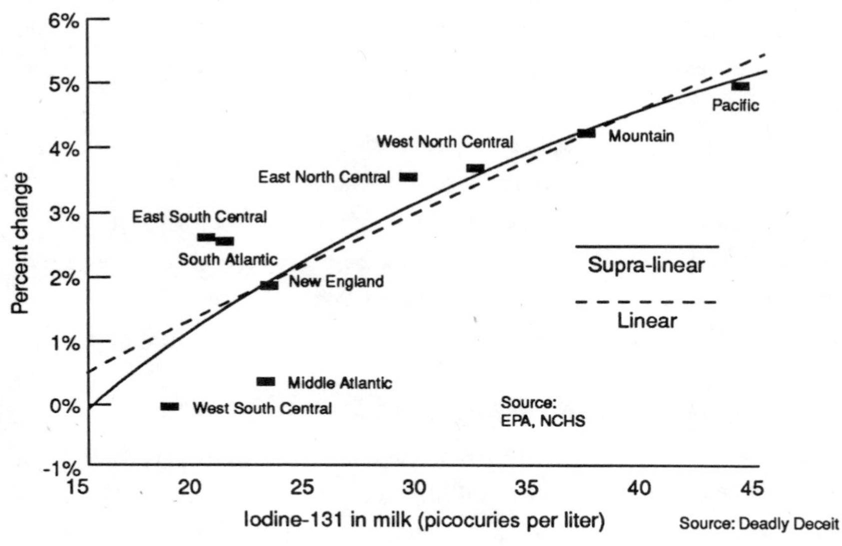

The statistical analysis shows that a supralinear, logarithmic curve (the continuous line in the data) fits significantly better than the linear relation (the broken line).

Such a logarithmic response curve is also characteristic of the Petkau Effect discussed earlier. The reason there were not more deaths in Europe, where the concentration in the milk was 100-1000 times higher, is because the supralinear response curve rises sharply only in the low dose region but then flattens out at high doses (see figure below). In simple terms, this means that while a minute amount of radiation will have a highly poisonous effect, twice that amount, for example, will not have twice the effect. Nevertheless, the damage done in Europe must have been greater than in the U.S.A., although the European populace was warned by the authorities not to drink the milk.

The figure shows the extrapolation of the logarithmic response curve from the previous graph to the higher iodine 131 concentration levels measured in Europe.[65,49] The straight line corresponds to the previously

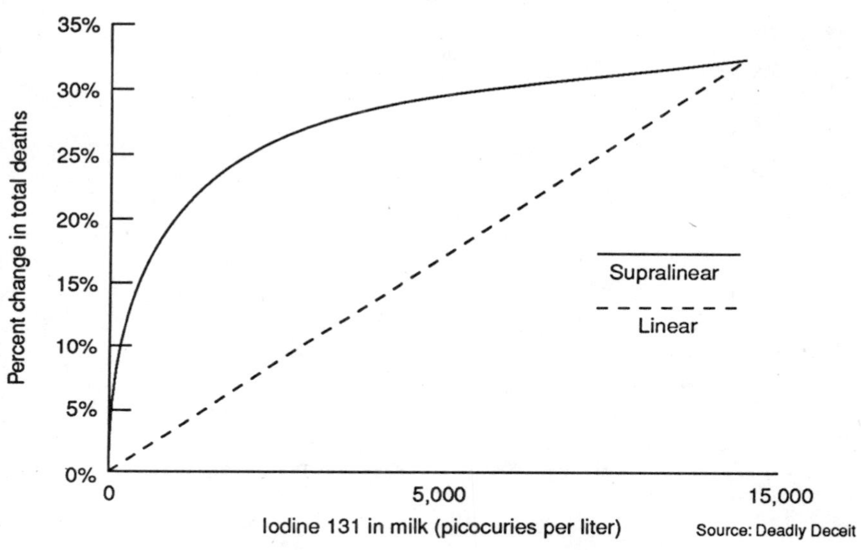

accepted, although incorrect linear response curve extrapolated from high to low doses. These Chernobyl data for the first time provide proof for a human population at the small doses resulting from fallout that the response curve is supralinear, i.e. that it is logarithmic or concave downward and not linear.[65,49]

The figure below shows the sharp increase in the infant mortality rate per thousand live births for the South-Atlantic states in the U.S.A. during the months of June and July 1986, in comparison with the corresponding months of the previous year. The drop in the birth rate here and in other regions was also shown to be a significant factor in the rise of the infant mortality rate, since this rate increases when the number of live births suddenly declines as a result of a rise in still-births.[65,49]

The victims of the Chernobyl accident fall predominantly into two groups: the elderly and the very young (including the developing fetus); and those suffering from life-threatening illnesses. One would also expect that the immune systems of those most affected were underdevel-

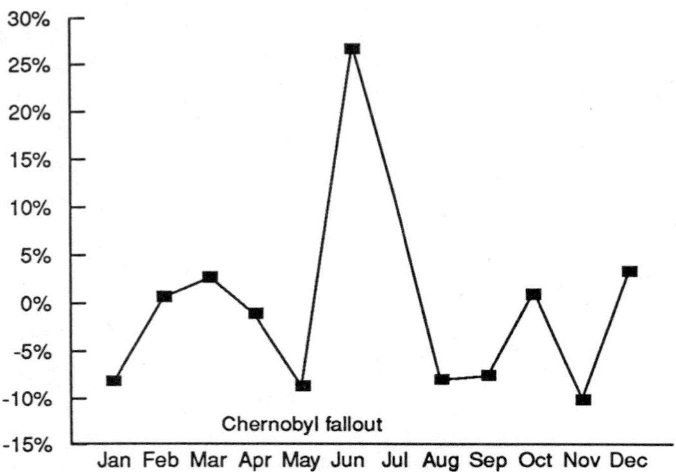

oped during the period of atmospheric atom bomb testing.[63] This hypothesis was later supported by the studies of Gould and Sternglass. According to their findings, compared with the average number of deaths of those over 65 years old in the summer months of 1983 to 1985, 7.4% more individuals in this age group died in 1986. This suggests that something must have accelerated their deaths. The number of deaths resulting from pneumonia was 18.1% higher for all age groups from May to August in 1986 compared with the same period in 1985. The deaths resulting from all infectious diseases combined rose 22.5% and those due to AIDS and associated infections 60.3%, supporting the hypothesis that damage to the immune system is the principal effect of fallout.

The chart below below shows the mortality changes grouped into four months in 1986 relative to 1985. For the first four months of 1986 these changes were much smaller than in the following summer months and dropped back down again during the last four months. As Gould and Benjamin Goldman put it in their book *Deadly Deceit,* from May to August 1986 there must have been a malignant force of mortality in action.

Gould and Sternglass found that not counting the deaths caused by accidents, suicide, drug abuse and violent crimes, the mortality rate in the 25-34-year age groups was some four times higher in 1986 than in 1985 than for the population as a whole. This age group represents those born in the 1950s, the decade of heaviest fallout from atmospheric bomb tests. As this group aged, the death rate resulting from their weakened immune system climbed more sharply than for those born before the A–bomb tests began. According to Gould and Sternglass, the probability that these increases in mortality could occur simultaneously by chance is less than one in a million.

Writing in *The Lancet,* Dr. Jens Scheer, Professor of Nuclear Physics at the University of Bremen, reported an increased death rate of newborn infants per thousand live births following the arrival of the Chernobyl fallout in the period from May 1986 to May 1987. (In southern Germany it amounted to 35%.)[66,67,29] Not surprisingly, his study showed that the regions most heavily exposed to radioactivity were also the ones showing the highest death rate. A critique of this study later published in *The Lancet,* was refuted in the same issue by Scheer and his group.[68]

The combination of extra deaths observed in the U.S.A. and Germany for humans, and land birds in California (see below) being mere coincidence is estimated by Gould at less than one in ten million.[49]

Mortality in the USA for 1986

Number of deaths resulting from:	Percent change relative to previous year January to April	May to August	September to December
Pneumonia	–5.7%	+18.1%	–3.4%
All infectious diseases	+2.3%	+22.5%	+15.7%
AIDS related infections	+11.6%	+60.3%	+19.8%

In 1989, Gould and Sternglass were able to present their research in a publication of the American Chemical Society. Their conclusions ran as follows:

The medical and scientific community has believed for a long time as a result of linear extrapolation from high doses, that low-level radiation from fallout and discharges from atomic installations can be dismissed because they represent a negligibly small danger. This is the principal assumption that scientists must now reevaluate. It underestimates the effect of low-level radiation on the most sensitive members of the population by a factor of approximately 1000.[65]

A Silent Summer in the U.S.A. in 1986

The Point Reyes Bird Observatory lies north of San Francisco. Beginning in 1975, resident ornithologist Dr. David DeSante studied 51 species of local land birds according to a carefully managed program. Every summer (May to August), the young birds were caught in standardized mist nets, then counted, weighed, measured, banded, and let go again.[70]

Thus, DeSante could demonstrate a significant correlation between the amount of winter rainfall (from November to March) and the success of breeding, i.e. the number of young birds caught the following summer. In the observation area, which has a temperate climate, 83% of the rainfall is in the winter and only five percent in the summer.

However, in the summer of 1986 DeSante became troubled by the dwindling number of young birds that were caught. "Instead of the two or three dozen birds that we had caught daily each year since 1975, there were at the most one dozen and sometimes only three or four."[67] On July 22nd—which would have normally yielded the highest bird activity—not a single young bird was caught in the net. The older birds were not singing anymore, "as if they had forgotten how to," said DeSante.[69] According to the winter rainfall, he had expected a normal breeding season. Instead, in 1986, there was such a low success in breeding as had never occurred before 1962, lying 62% below the ten year average).[70]

DeSante made inquiries of colleagues in California as well as in Washington and Oregon.[67] The outcome of his investigations was that an identical or similar picture prevailed everywhere that rain had fallen on

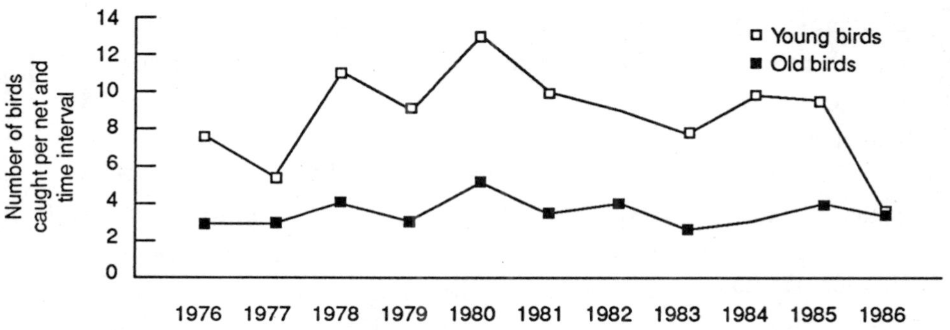

May 6th. The radioactive cloud from Chernobyl had passed over these areas on that day. Radioactivity is brought down from the atmosphere by rain. Where no rain had fallen, there was no drop in the number of new birds.

Climatic factors were excluded as the cause of the decline, as were migratory birds that had spent the whole of the winter elsewhere, returned and had had a poor breeding season. So were the use of pesticides, herbicides or hunting by people or birds of prey in this protected wildlife preserve. The only explanation which remained was the radioactive cloud from Chernobyl and the accompanying rain.[70]

A study of the food chain substantially supported the Chernobyl hypothesis. It showed that the drop in the number of young birds was independent of migratory behavior, territorial habits, location of nests, etc., but not independent of the feeding conditions of the young birds. The smallest decline in the young bird population was seen amongst swallows and goldfinches, which breed rather late (iodine 131 decays rapidly; its half-life is eight days). Woodpeckers that feed their young mainly on insects were completely unaffected: the insects they eat in turn feed on dead or decaying wood that does not absorb rain and therefore iodine 131. The other land birds had eaten insects contaminated by fallout that had precipitated in affected areas.

The effects of iodine 131 have been demonstrated in animals and humans. It concentrates in the thyroid gland where, among other things, it interferes with the production of vital hormones.[69,65] DeSante cited the Petkau Effect as the explanation why such small doses could produce such a large effect. The same effect can be found in birds, where chronic doses of gamma rays (i.e. small dose rates) are much more dangerous for the growth process than the same dose from a single X-ray exposure. This was demonstrated in a study carried out by the Canadian Atomic Research laboratory in Manitoba.[71]

Following the publication of an article in *The Condor,* DeSante was criticized. Atomic energy advocates pointed out that autopsies performed on the young birds revealed no detectable amounts of cesium 137 or strontium 90. However, DeSante showed that on the basis of measurements of the rain by the Environmental Protection Agency, there had been high concentrations of iodine 131 (up to 2,000 pCi) and at the same time little or no cesium or strontium.[69] With this evidence, DeSante was able to show that iodine 131 by itself correlated with the iodine 131 concentration in the milk, using EPA measurements. A similar picture was found to exist across the United States on the basis of bird counts from the "Breeding Bird Survey" sponsored by the U.S. Fish and Wildlife Service.[49] However, in Switzerland, according to a study by Noordwijk, no bird deaths were found as a result of the effects of Chernobyl. DeSante attributed this to the late breeding season in Switzerland due to its colder climate, combined with the fact that the half-life of iodine 131 is only eight days.[72]

Disposal of Atomic Waste Remains Unsolved Worldwide

The problem of atomic waste disposal has remained virtually insoluble. Professor Alexander Tollmann, chairman of the geology department at the University of Vienna, has also reached this conclusion:

> Geology is not a precise and predictive science, and the hubris of technology founders on its probability calculations. Deadly radioactive wastes must be kept away from the biosphere for 100,000 to 1,000,000 years. For waste dumps containing such massive amounts of long-lasting alpha activity, long-term reliable forecasts concerning the location, strength, and variability of earthquakes must be available into the distant future, as well as forecasts of changes over a period of 10,000 or more years,

> together with the consequences of ice-ages and patterns of drought and rain. As a geologist one therefore has the moral duty to initiate a search for permanent disposal sites only after the last atomic installation is closed down. Otherwise, every glimmer of hope would be misused by the propaganda machine of the nuclear establishment to give false promises of safety when this is far from being in sight.[73]

With each day of atomic power production, the mountain of waste grows for us, our children, and future generations as a deadly inheritance. It is therefore understandable that much of the public rejects the search for permanent nuclear waste dumps in areas where they live. Moreover, concerned citizens and groups have taken steps to heighten the public's awareness of the danger of atomic energy. Environmental organizations such as Greenpeace demand that the waste be stored in retrievable intermediate storage, preferably on the sites of nuclear installations themselves. In this way, the public would have waste from this ill-conceived form of energy production always "in view," which is, of course, exactly what the atomic-energy advocates want to avoid. They would prefer to "bury" it for all time, no matter what the biological cost to future generations. In retrievable types of storage, nuclear waste is at least temporarily under control. In the event of leakage or inflow of water, counter-measures can be taken. However, even here the risk is not totally eliminated.

The Greenhouse Effect

Climatic research has long been concerned with the possible warming of the earth. It is feared that with the rising combustion of fossil fuels, the carbon dioxide produced may increase its atmospheric concentration and raise the likelihood of significant global warming. According to Greenpeace, the prospect of a climatic catastrophe is being misused by the atomic energy industry, harassed after Chernobyl, to push for the replacement of fossil fuels by nuclear energy.[74]

In a press conference held on September 13, 1989, the Swiss Physical Society (SPG) stated that the Greenhouse Effect was the only plausible explanation for the half-degree (centigrade) warming of the Earth over the last 100 years. The SPG is supported by a study of the Paul Scherrer Institute. Therefore, the Physical Society recommended, among other steps, an increase in the production of nuclear energy.[75,83]

This recommendation is not adequately supported by current research into the Greenhouse Effect, as *Science* magazine reported, in its coverage of a symposium of leading climatologists held in Amherst, Massachusetts.[76,83] At this 1989 meeting, the shortcomings of the existing climate models were discussed. Among other things, the uncertainties surrounding global warming were related to the behavior of the oceans with respect to heat transfer, the role of clouds, the influence of solar activity, and volcanic eruptions, since volcanic ash in the stratosphere leads to a cooling of the lower atmosphere.[76,77,78] As an example, according to the George C. Marshall Institute in Washington, the curve showing the temperature increase over the last century correlates better with that for solar activity than with the CO_2 increase. Moreover, the form of the curve showing solar activity over the past millenium actually suggests the possibility of an impending cooling of the earth in the next century.[79,80]

As a result of the controversy surrounding global warming, the symposium ended inconclusively. At the closing press conference, the following was nevertheless agreed upon by all 40 participants: "... That it is indeed tempting to attribute the half-degree centigrade global warming over the past 100 years to the Greenhouse Effect. Because of natural variation in temperature, such a conclusion might also not have the slightest credibility." This statement must be considered representative of all present Greenhouse research, since, according to *Science,* there are—insofar as climatic modelling and computers play a role—five groups of Greenhouse Effect researchers which can be considered world class, and they were all represented at the Amherst symposium.[80]

Disregarding the unacceptable dangers of radioactivity, nuclear energy could not solve the problem of the Greenhouse Effect. For example, if we had wanted to halve oil consumption in 1986, about 7,400 atomic power stations would have been required.[81] The CO_2 content of the atmosphere in the pre-Industrial Age amounted to 280 ppm (parts per million) compared with 340 ppm today. Yet, over the past 30 years, the 375 atomic power stations in operation in 1986 had prevented a mere two ppm of CO_2 from entering the atmosphere.[81] CO_2 produces only half the total Greenhouse Effect. The other half comes from trace gases such as methane, nitrous oxide, and chlorofluorocarbons (CFCs). CFCs could and ought to be banned immediately. In addition, they contribute to the destruction of the ozone layer. Steam, aerosol and radioactive krypton 85 are also under discussion as possible Greenhouse gases. Radioactive krypton is released in the daily operation of nuclear plants and in still larger amounts through the reprocessing of burned-out nuclear reactor fuel elements. The amount of this isotope entering the atmosphere is climbing at a rate of five percent per annum, and it has already affected the distribution of electric fields in the atmosphere, as discussed earlier.[57,82] The release of radioactive gases alone changes the frequency and distribution of global precipitation (Kollert/Butzin).[82]

Previously, imminent climatic catastrophe was often taken for granted.[76,83] Recently, the theme has been treated with greater caution, and the "doomsday" conclusion regarded as more controversial (See *Science* [November 24, 1989], *Energie und Umwelt* [March, 1988], *Science* [December 1, 1989], *PM Magazin* [December, 1989], *Neue Zürcher Zeitung* [January 27/28, 1990].[78,81,84,85,86] A change in sentiment was also expressed in a conference on the global climate convened in April, 1990 by President Bush.[87] According to a recent *Science* article, satellite measurements will be required for another decade to furnish reliable proof of a possible global warming.[101] This does not mean that the problem is not to be taken seriously, since the average temperature has indeed risen in recent years—although the increase is still within the range of natural variation and may reflect a "rebound" from an abnormally low temperature during the period of large-scale atmospheric bomb testing.

The reduction in CO_2 emissions can be more safely addressed by the development of renewable energy production technologies such as small-scale hydroelectric power stations (about 3,500 of which have been lying dormant in Switzerland alone since 1928) or the utilization of waste heat from existing fossil-fueled electric power generating stations. The technology is basically ready, and the energy yield would be sufficient to cover every demand that, for example, Switzerland requires, assuming that known conservation techniques (in particular improved efficiency) are applied.[88-92] According to a Swiss government study, society could wean itself from dependence on nuclear energy in a relatively short time and without any loss of living standard (EGES-Report).[91]

Worldwide, these proposals for environmentally benign energy production must be encouraged through political means, i.e. by giving them a fair chance of success. This requires that large amounts of money should not be invested in nuclear energy research and development but instead directed to the areas of increasing energy efficiency and the development of renewable sources. The prerequisite is clearly a phase-out of nuclear energy.

The New BEIR V report

A new report on the biological effects of ionizing nuclear radiation (BEIR V), prepared by the National Academy of Sciences, was published in February 1990. Despite many contradictions and the withholding of important facts dealing with supralinearity, membrane damage, the Petkau Effect and a number of important epidemiological studies, it could signal the ultimate end for nuclear energy.

The BEIR V report estimates a 300% higher cancer risk compared to the 1980 BEIR report, as a result of revisions in the earlier Hiroshima/ Nagasaki dosimetry. The earlier analysis had used a linear quadratic model with an additional reduction factor in the lowest dose regions. The new report estimates that as many as 800 additional cancer deaths per million persons would be produced by a dose of 100mSv, or one rem.[93,9] Never before has a leading official organization estimated such a high risk.

A reduction factor in the lowest dose regions is no longer used, and a purely linear model like the one Radford had already demanded in 1980 was employed. (See the earlier discussion of "Intrigues and Manipulation of Science.") It is now admitted that the risk at very small doses could be either larger or smaller.[79]

However, in order to protect the nuclear industry, the supralinear model which leads to a higher estimated risk is not discussed at all, and thereby effectively covered up. The President of the BEIR V committee, Professor Arthur C. Upton, suggested in a press conference dealing with the new report that the permitted annual maximum level need not necessarily be reduced, although perhaps this might be necessary for accumulated lifetime doses.[93] Despite these efforts to hide the importance of the new findings, their significance is becoming more and more evident. For the first time, one of the leading establishment organizations dealing with the question of radiation "violated" the expected pattern of minimizing the dangers of radiation.

• The report states that it is no longer possible to argue that the estimated small releases of radiation in the United States, due to nuclear plant operation and weapons testing, is of no consequence when compared with natural and medical sources. On the contrary, the BEIR V report questions the validity of this argument because of the uncertainties in the dosimetry and equivalent dose, and does not adopt the dose estimates for the various sources published by the National Council on Radiation Protection and Measurements.[6]

• Furthermore, the report acknowledges the decisive role of free radicals. According to the BEIR V report, for ionizing radiation in the low dose range, such as X-rays and beta rays, two-thirds of all biological damage is caused by the indirect effects of free radicals.[7] All this is equivalent to a revolution in radiation safeguards. But it is of interest to note that Petkau and membrane damage were no longer referred to, in contrast to the detailed discussion they were accorded in the BEIR III report of 1980.

• For the first time, a whole series of studies was discussed that shows an increased risk to the public as a result of high natural radiation levels, diagnostic X-rays, fallout from A-bomb explosions, nuclear accidents and exposures involving radiation workers at doses below those presently allowed. And for the first time, these findings have not been rejected out-of-hand just because they cannot be understood in terms of the Nagasaki/Hiroshima risk estimates.

According to BEIR V, they could, on the contrary, raise legitimate questions about currently accepted estimates due to the problem of extrapolation from high to low doses.[8] Nevertheless, the BEIR V report tried to discredit the overwhelming evidence for serious low dose effects. Yet the data published in the report show that the dose response curves in the low dose range need not necessarily be quadratic or linear but could be supralinear, as would be consistent with the Petkau Effect.

Deadly Deceit

Deadly Deceit is an easily comprehensible book by Dr. Jay M. Gould and Benjamin Goldman and has enormous significance.[49] The lucid examination of the harmful effects of nuclear accidents given by the authors provides significant support for the hypothesis that millions of Americans have died as a result of the releases from weapons tests and nuclear installations. For example, following the accidents at Savannah River and Three Mile Island, there were significantly more than the expected number of deaths due to all causes, while after Chernobyl there was a sharp decline in live births, a rise in infant mortality and an increase in deaths due to all causes combined, in adults as well as infants. The book presents clear evidence that an attempt was made to hide the facts from the general public by both state and federal agencies.

The book also examines the new BEIR V report. It supports in a striking manner the results of the present volume. The authors specifically hold the Petkau Effect to be responsible for biological damage at low dose rates. The assertions made in Gould and Goldman's book are strongly supported and corroborated by the many studies cited in the BEIR V report.

Until recently, those who have been responsible for radiation-protection of workers and the public have set and maintained high maximum permissible levels. What these officials refuse to admit, is that in order to fully protect both human and animal life, maximum permissible levels must be set at zero. The production and discharge of artificial radioactivity into our environment must be forbidden on principle. Ionizing radiation should only be used on an individual basis, and radiation exposure accepted only voluntarily, and when it is ethically justifiable as, for example, in medicine and in special areas of industry and research.

Professor John Gofman, who had taken part in the development of the A-bomb and was the first director of the Department of Biomedicine at the Lawrence Livermore Laboratory, had already sharply challenged his scientific colleagues as early as 1979, with the following statement:

> There is no way I can justify my failure to help sound an alarm over these activities many years sooner than I did. I feel that at least several scientists trained in the biomedical aspects of atomic energy—myself definitely included—are candidates for a Nuremburg-like trial for crimes against humanity through our gross negligence and irresponsibility. Now that we know the hazard of low-level radiation, the crime is not experimentation—it's murder.[94,95]

Today, an increasing number of radiation experts and other scientists oppose the official radiation standards. Thus, in Germany, an "Association for Radiation-Protection" was established by a number of noted professors from various German universities. In an address at the opening meeting in March 1990, Professor Inge Shmitz-Feuerhake of the University of Bremen said the following:

> The revision of the traditional opinions concerning the harmlessness of low doses of ionizing radiation cannot be satisfactorily achieved by the present scientific bodies. Under the influence of nuclear power propaganda, the established associations have avoided every critical analysis of the data or stood clear of the debate entirely. Among other things, the tragic result of this policy has been that many radiation victims were and are still being produced which could have been avoided by an objective examination and evaluation of existing scientific investigations and the subsequent incorporation of the findings into radiation-protection measures.

The international radiation-protection authorities only issue recommendations. No nation is forced to adhere to these guidelines. Yet, despite comprehensive documentary material, authorities in countries around the world have underestimated the effects of small doses. Thus, it is not unreasonable to question whether the responsibility for the protection of the public in the area of radiation is being taken seriously by the agencies charged with this duty.

Switzerland, paradigm of an orderly society with every technological resource at the authorities' disposal, is a good case study of how government evades dealing with the consequences of nuclear radiation. After a parliamentary investigation in 1990 shed light on the inertia, incompetence and self-serving attitudes in the ministry of justice and the ministry of defense, it would hardly be surprising if arrogance and high-handedness were discovered to play a role in the bureaucracy responsible for public health and radiation safety. Decisions and recommendations favor nuclear energy time and again at the expense of the public. This situation appears to exist in the case of radiation-protection in every country, and when criticism arises, the most vocal protestors—who are exercising their legal and democratic rights—are kept under police surveillance as "Enemies of the State."

"Atoms for peace" and "atoms for war" have always belonged together. One program is contingent on the other. Without the powerful incentive of atomic weapons, the investments and sacrifices for the development of nuclear energy carried out under the most rigid secrecy would not have been possible.

This has all been stated in the new "Greenpeace Handbook of the Atomic Age."[74] This carefully documented work accurately describes the ways in which the development and application of nuclear energy, as well as weapons systems, continues to be paved with unforeseeable accidents and disasters. Experts in many nations carried out the research for Greenpeace, and they present heretofore secret information. The book also contains a large number of documents, diagrams and photographic material. The U.S.A., Soviet Union, Great Britain, France, and China will not remain the sole atomic powers for long. Today India, Israel, Pakistan, Brazil and South Africa are in a position to manufacture nuclear weapons. The ability to do this has been acquired under the guise of "peaceful" atomic programs.[7]

More countries will follow: already Iraq, North Korea and Libya are developing nuclear weapons.

Whoever supports nuclear energy supports the production of nuclear weapons. Sadly, the possibility of renewable energy production is too often overlooked. Easily harnessed, such energy is feasible, safe, and, assuming its use would replace nuclear energy, would save lives. Yet the necessary conditions to achieve this goal can only be brought about by a firm decision to reject nuclear energy.

50,000 to 100,000 Deaths

The Swiss Federal Health Ministry (BAG-Bundesamt für Gesundheitswesen) in its Bulletin No. 20 of May 28, 1990, reported that Sternglass and myself maintained that emissions from nuclear power stations could be very dangerous.[96] In an effort to refute this claim, the Ministry accused Sternglass of using false data to demonstrate an increase in infant mortality near the Three Mile Island plant. But even the Secretary of Health of Pennsylvania at the time, Dr. Gordon Macleod (who afterwards was forced to resign), acknowledged the sharp increase in infant mortality in a five to ten mile radius of the reactor, in agreement with the independent investigation of Sternglass in the Harrisburg and Pittsburgh hospitals.[49,97] These facts support Sternglass' earlier findings, which showed that there was a higher infant mortality during the summer months, corresponding to the direction of wind in the first days of the accident. Thus, the effect was not only limited to the proximity of the reactor.[97]

Very recently, Gould and Goldman documented in *Deadly Deceit* that the original monthly data on infant deaths in the *U.S. Monthly Vital Statistics* were altered in order to remove embarrassing peak values not only after the TMI accident, but also in other earlier accidents. Similarly, the two melt-downs of fuel rods in 1970 and the accompanying emission at the Savannah River Plant (SRP), a nuclear weapons production facility in South Carolina, were covered-up at the highest government levels. These accidents were admitted to have taken place by the government for the first time in 1988, in the course of Congressional hearings.

The data in both cases were nevertheless poorly "trimmed." Thus, Gould and Goldman could establish a statistically significant 50,000 to 100,000 increase in deaths for SRP, and TMI, for the latter up to 500 miles away. In the case of the county where TMI was located (Dauphin County), a highly significant increase in the infant mortality rate was found, corroborating Sternglass' earlier finding. The statistics analyzed by Gould and Goldman confirm those of Sternglass.

Aside from trying to discredit Sternglass' findings around TMI, the Swiss Ministry of Health attempted to dispute the existence of the inverse dose rate effect (the Petkau Effect), thereby questioning the supralinear dose response curve in living systems. The Ministry's article claims that Petkau himself had given evidence contradicting the significance of his findings. According to the Ministry, Petkau cited other experiments that showed that at high dose rates, the molecules of a phospholipid membrane are strongly cross-linked, and would therefore be more protected against radiation damage than at low dose rates. The Ministry neglects to mention that its study is based on papers written 20 years ago. According to the Ministry, a cross-linking in living membranes would be dangerous, so Petkau's experiments are only valid for artificial membranes. And the article fails to note that the effect has long since been observed by others, not only in artificial membranes but also in mouse liver cells and human lymphocytes (white blood cells).[48,46,4] As long ago as 1980, the authors of the BEIR III report (on pages 463-469) admitted the potential importance of the Petkau Effect. Finally, the reader is referred to the personal communication of Petkau cited above. Petkau's discovery deserves nothing short of the Nobel Prize. But it was Sternglass who, in 1974, first pointed out the significance of the Petkau Effect for understanding the serious consequences of fallout on humans.

Nonetheless, as recently as 1989, the radiation expert Dr. E.H. Paretzge of the Association for Radiation-Protection and Environmental Protection in Neuherberg (Münich) characterized the Petkau Effect as insignificant, in an article published in the journal *Physikalische Blätter* (Vol. 45, pp. 16-24). Paretzge's argument was refuted in a joint article by Professor E. Lengfelder of the Radiation Biology Institute, University of Münich, and Petkau. In their reply, according to the *Süddeutscher Zeitung* of December 19, 1989, they stated that Paretzge ignored evidence that invalidated his argument and instead referred to vague descriptions of unpublished animal experiments that supposedly did not show the inverse dose-response. According to the newspaper report, Lengfelder found himself in agreement with many professional colleagues, who shared his view of the seriousness of Petkau's findings.

Paretzge maintained that responsibility for providing scientific proof rested on those who have proceeded from the assumption that small radiation doses and small dose rates are a great danger to human health. I recently learned that exactly the kind of scientific proof Paretzge called for—of the seriousness of low dose exposures for the immune system—already exists. It has been ignored, and effectively covered-up, by the authorities because of the devastating implications for the nuclear industry.

As published in the highly-regarded journal *Radiation Research,* this study was funded by the U.S. Department of Energy, and carried out at the Ecological Sciences Department of the Battelle Northwest Laboratory at Hanford, Washington. It found a permanent suppression of immune system function in rainbow trout exposed to extremely low concentrations of tritium during the first twenty days of embryonic development at doses as low as 40 millirads, or 0.4 mSv.[98] Moreover, the decline in immune system capability was related logarithmically and not linearly to the dose. This pattern was what Petkau reported for a free-radical type of action on cell membranes rising rapidly at very low doses and levelling off at the higher doses—just as for total mortality rates after Chernobyl, and cancer rates around Three Mile Island.

The authors of this report were fully aware of the significance of their findings for the future of the nuclear industry. In their words:

> This study is of practical importance because tritium is, and likely will continue to be, a major contributor to radioactivities in effluents from nuclear power plants, both fission and fusion designs, and other facilities. The radionuclide is not easily removed by conventional waste treatment practices; thus there is increasing concern as to the ultimate concentration in and effect upon biological systems. Furthermore, any likelihood of tritium having radiobiological effects in the range of the maximum permissible concentration (MPC) of 0.1 microCuries per milliliter justifies careful attention.

The experiments showed statistically significant effects at concentrations as low as 0.009 and 0.08 microCuries per milliliter, well below the maximum permissible concentrations. But radiation-protection and public health authorities all over the world chose to ignore these warnings, and instead raised the permissible concentrations of many radioactive isotopes in the water we drink, the food we eat, and the air we breathe.

The outdated arguments of the pro-nuclear forces are also refuted by a series of very recent studies published in the U.S.A.:

1. Pilgrim reactor

For the first time, a study in the vicinity of a nuclear power plant, prepared by a State Public Health Department, reported a significant increase in the leukemia rate.

In October 1990, the Massachusetts Health Department found that for people who live and work in 22 towns located roughly 20 miles from the Pilgrim nuclear power plant near Plymouth, the incidence of leukemia was four times higher than normally expected.[99] Governor Michael Dukakis then demanded a reduction in the atmospheric radioactive releases.

2. Oak Ridge National Laboratory

A comprehensive study by scientists at the University of North Carolina, led by S. Wing, investigated the effects of low dose exposures on 8,313 men who worked in the plant for more than 30 days over a period of 24 years (1943-1972). The results, published in the *Journal of the American Medical Association* (JAMA) in March 1991 (Vol. 265, page 1397), showed that the cancer risk of the employees was ten times higher than would have been expected from the Hiroshima/Nagasaki data. This risk estimate for the Japanese A-bomb survivors is, of course, based on single exposure at high doses and dose rates due to external radiation from a nuclear explosion.

3. Three Mile Island Reactor

A study conducted by Maureen C. Hatch and her colleagues at the Columbia University School of Public Health in New York, published in *The American Journal of Epidemiology* (Vol. 132 September 1990, page 397), claimed to have found no increased cancer risks from the 1979 accident at the TMI reactor. However, in the second edition of *Deadly Deceit,* using the data from the Hatch report, Dr. Gould drew exactly the opposite conclusion: He found a significant relation between the dose received and the Standard Incidence Ratios after the accident. Moreover, the fact that the relation had a supralinear or logarithmic form again supports the role of the Petkau Effect.[49]

4. Trojan Reactor

The second edition of *Deadly Deceit* reported a study by Gould and Sternglass around the Trojan Reactor near Portland, Oregon. Their investigation concluded that increased leukemia and cancer rates in the nearby counties appear to be related to the radioactive emissions of this reactor, following damage to its nuclear fuel rods. Again, a supralinear dose-effect curve was found, which the authors attribute to immune system damage produced by the Petkau Effect.

5. Study of the National Cancer Institute (NCI)

This study, published in 1990 ("Cancer in Inhabitants Living Near Nuclear Facilities"), concluded that living near nuclear facilities did not increase the chance of cancer. Sixty-seven American nuclear facilities of all types were investigated. The utility operating the Trojan reactor tried to refute the Gould/Sternglass findings by citing the NCI study. However, in *Deadly Deceit,* the authors point out the many built-in biases of the NCI report, and in fact, they support their findings pertaining to the Trojan reactor with the NCI data for this reactor.

The New ICRP Recommendations

The latest ICRP publications do not reflect the consequences of new radiobiological findings.[100,26] For example, the large number of epidemiological studies which indicate that supralinear dose-effect curves hold in the low dose rate case of chronic exposures are not taken into consideration. Again, the new permissible doses do not reflect a primary concern for the protection of human life but are designed to permit the continued operation of nuclear reactors.

The commission estimates, according to the revision of the Hiroshima-Nagasaki data, that the risk of cancer deaths is 3.2 to four times higher than was assumed in the past.[26] As early as 1977, the cancer risk was considered to be very significant compared with 1958, when commercial nuclear energy was first introduced. In 1958, nearly all scientists and the general public believed that there was no cancer risk produced by the use of nuclear energy.

Despite the evidence for a significant risk, especially for the developing infant in utero, the ICRP, in an effort to protect the nuclear industry, did not reduce the permissible levels in 1977. Therefore, the permissible dose for occupational exposures was kept at 50 mSv (5 rem) per year.

The table below shows the cancer risks assumed by the ICRP in the years 1958, 1977 and 1990 (number of cancer deaths per million people, each exposed to a single dose of radiation of 10 mSv [1 rem]).

Year	Maximum Permissible Dose	Number of Cancer deaths
1958	5 rem	0
1977	5 rem	125
1990	5 rem	125
1990 (Draft)	5 rem (2 rem av.)	400-500

The table also shows that from 1958 to 1990, although the maximum permissible occupational dose remained constant with the new draft proposal, the cancer risk rose from 0 to 125. Therefore, the permissible doses ought to have been lowered 125 times. Incredibly, the latest reduction adopted in the 1990 draft was only 2.5 times, despite the much larger increase in risk estimates.

The new permissible dose for occupational exposure is fixed at an average valve of 20 mSv (2 rem) per year. This average dose is actually based on a maximum valve of 100 mSv (10 rem) in any five year period and in any single year a dose as large as 50 mSv (5 rem) is still allowed. For the general population, the commission has decided on a permissible dose of 1 mSv (100 millirem) per year averaged over a five-year period.

The new cancer risk factors of the ICRP 60 (1990) are still much too low, since a reduction factor of two is applied for the low dose range below 200 mSv or 20 rem. Thus, the predicted number of cancer deaths calculated on the basis of the high dose and dose rates are further reduced in the latest ICRP report, despite the new epidemiological studies cited in the BEIR V report.

Moreover, the new maximum permissible doses were set in a reckless manner. "It is the commission's intention that dose limits are set such that continued exposure just above the dose limits would result in added risks that legitimately could be called 'unacceptable.' The dose limits therefore are set just at the dividing line between what is barely tolerable and what is intolerable."[26] However, what is supposed to be barely tolerable is decided arbitrarily by the ICRP, and quite obviously the decision was made only in the interest of the nuclear industry.

The ICRP is continually caught up in its own web of increasingly complicated theoretical models. The commission's work proves that the widespread radioactive contamination produced by the nuclear industry can never be brought under control.

For example, the number of organs for which weighting factors are given were increased from seven to thirteen. This, too, is inadequate. As long ago as 1969, the ICRP already mentioned 27 different critical organs. Numerous new indices of the consequences of exposure are now created: attributable risk of death, potential life-time lost, annual risk of death and the probability of dying in any year.[26]

Today, we must aim to eliminate all known carcinogens, including fission products, and to discontinue their production and deliberate release into our environment. Complex, irrelevant calculations of the sort put forth by proponents of nuclear energy will not shield us from the effects of radioactive releases.

References

Parts I and II

1. Arbeitsgruppe Wiederaufbereitung, University of Bremen, *Atommüll,* Rowohlt Taschenbuch, 1977.
2. V. E. Archer, "Geomagnetism, Cancer, Weather and Cosmic Radiation." *Health Physics,* 34, 1978, p. 237–247.
3. R. Alvarez, "Radiation Standards and A-bomb Survivors." *Bulletin of the Atomic Scientist,* Oct. 1984, p. 26–28.
4. M. Barcinski et al., "Cytogenetic Studies in Brazilian Populations exposed to Natural and Industrial Radioactive Contamination." *American Journal of Human Genetics* 27, 1975, p. 802.
5. *Basler Zeitung,* "Bundesrepublik verbietet Rheumamittel." Jan. 27, 1984.
6. *Beaver County (Pa.) Times,* "State Panel Questions Radiation Safety." June 7 1974.
7. BEIR 1972, p. 2, 18.
8. BEIR 1972, p. 22.
9. BEIR 1972, p. 44.
10. BEIR 1972, p. 45.
11. BEIR 1972, p. 46.
12. BEIR 1972, p. 48.
13. BEIR 1972, p. 56,57.
14. BEIR 1972, p. 58.
15. BEIR 1972, p. 62.
16. BEIR 1972, p. 69, 70.
17. BEIR 1972, Chapter VII, p. 83.
18. BEIR 1972, p. 90.
19. BEIR 1972, p. 91.
20. BEIR III, 1980, p. 5.
21. BEIR 1980, p. 72.
22. BEIR III, 1980, p. 72.

23. BEIR III, 1980, p. 96.
24. BEIR III, 1980, p. 98.
25. BEIR III, 1980, p. 110.
26. BEIR III, 1980, p. 180.
26. BEIR III, 1980, p. 180.
27. BEIR III, 1980, p. 193.
28. BEIR III, 1980, p. 243.
29. BEIR III, 1980, p. 31, 244.
30. BEIR III, 1980, p. 245.
31. BEIR III, 1980, (E. P. Radford) p. 227–253, (H. Rossi) p. 254–260.
32. BEIR III, 1980, p. 463–469.
33. J. Bleck and I. Schmitz-Feuerhake, "Die Wirkung Ionisierender Strahlung auf die Menschen." University of Bremen, Sales No. K 012, 1979.
34. W.L. Boeck, "Meteorological Consequences of Atmospheric Krypton-85. *Science,* 193, July 16 1976, p. 195–197.
35. H. Brunner, *Die sanften Mörder: Warnruf oder Schauermärchen¿ Tages Nachrichten,* Bern, May 6 1972.
36. H. Brunner. "Buchbesprechung: 'Die sanften Morder: Atomkraftwerke demaskiert'." *Broschüre des Sekretärs des Fachverbandes für Strahlenschutz,* Zürich, 1972, p. 2, 12.
37. J.B. Bucher. "Bemerkungen zum Waldsterben und Umwelschutz in der Schweiz." *Forstwissenschaftliches Centralblatt,* April 1984, p. 23, 24.
38. *Bulletin of the Atomic Scientists,* "The Deepest Hole in the World." June 29 1984, p. 1420.
39. *Bulletin of the Atomic Scientists,* "First Look at the Deepest Hole." Sep. 29 1984, p. 1461.
40. West German Federal Ministry of the Interior, *Strahlenschutz-Forschungsbericht 1982,* St.sch. 812, Gesellschaft fur Reaktorsicherheit (GRS), Cologne.
41. West German Federal Ministry of the Interior, Radiologische Langzeituntersuchungen in bayerischen Oberflächengewässern." *Strahlenschutz-Forschungsbericht 1982,* St.Sch. 501 GRS, Cologne.
42. M. Burri, "Überforderte Geologen¿" *Basler Zeitung,* August 6 1981.
43. W.S. Chelack, M.P. Forsyth, A. Petkau, "Radiobiological Properties of Acholeplasma Laidlawii B." *Canadian Journal of Microbiology* 20, 1974, pp. 307-320.
44. Coasta-Ribeiro et al., "Radiological Aspects and Radiation Level Associated with Milling of Monazite Sands." *Health Physics,* 28, 1975, p. 225.
45. M. De Groot, "Statistical Studies of the Effect of Low-Level Radiation from Nuclear Reactors on Human Health." *Proceedings of the 6th Berkeley Symposium,* July, 1971, University of California Press.

46. Dertinger et al., *Molekulare Strahlenbiologie.* Heidelberger Taschenbucher No. 57/58, Springer Verlag 1969, p. 4.

47. G. Drake, *A Report on Selected Charlevoix Country Statistics for the Aliquipa Hearings.* July 31 1973.

48. Eidg. Expertengruppe Dosiswirkung, *Wirkungen kleiner Strahlendosen auf die Beyölkerung.* Report of June 1981, p. 73.

49. R.W. Field et al., "Iodine-131 in Thyroids of the Meadow Vole Microtus Pennsylvanicus in the Vicinity of the Three Mile Island Nuclear Generating Plant." *Health Physics,* Vol. 41, August 1981, p. 297–301.

50. *Frankfurter Allg. Zeitung* "Däs Risiko natürlicher und kunstlicher Strahlung." Hans Zettler, May 23 1984.

51. I. Fridovich, "The Biology of Oxygen Radicals," *Science,* 201, 1978, p. 875.

52. H. Fritz-Nigli, *Strahlengefährdung, Strahlenschutz,* Werlag Hans Huber, Bern 1975, p. 81, 212, 211, 210, 208.

53. H. Fritz-Niggli, "Presentation, SVA-Informational Meeting," 22.23.11., Zurich-Ierlikon, 1982.

54. H. Fritz-Niggli, "Problematik von Risikoschätzungen," *SVA-Bulletin,* No. 13, 1983.

55. J.T. Gentry et al., "An epidemiological Study of Congenital Malformations in New York State." *American Journal of Public Health,* 49, 1959, p. 497-513.

56. Global 2000, *Bericht des Präsidenten der USA* Verlag Zweitausendeins, Postfach, 6000 Frankfurt 61, p. 426, 427.

57. J.W. Gofman and A.R. Tamplin, *Report of the Investigations Committee on Air and Water Pollution in Public Works.* US Senate, 91st Congress, Nov. 18 1969.

58. J.W. Gofman and A.R. Tamplin, *Congress Seminar,* Radiation Laboratory of the University of California, Berkeley, April 7-8, 1970.

59. J.W. Gofman and A.R. Tamplin, *Populations Control through Nuclear Pollution,* Nelson Hall Comp., Chicago 1970.

60. J.W. Gofman, *Radiation and Human Health* Sierra Club Books, San Francisco, 1981.

61. R. Graeub, *Die sanften Mörder: Atomkraftwerke demaskiert,* Albert Muller Verlag, 1972, Rüschlikon-Zürich, Paperback edition: Fischer Taschenbuchverlag. Frankfurt, 1974.

62. L.H. Hampelmann, *Lancet,* 1983, p. 273.

63. H.P. Hanni, "Ist künstlich erzeugte mit natürlicher Radioaktivität vergleichbar¿" *Basler Zeitung,* May 13 1981.

64. J.M. Harrison, "Disposal of Radioactive Waste." *Science,* 226, Oct. 5 1984, p. 11-14.

65. ICRP Publication No. 8, 1966, p. 2.

66. ICRP Publication No. 8, 1966, p. 8.

67. ICRP Publication No. 8, 1966, p. 56.
68. ICRP Publication No. 8, 1966, p. 60.
69. ICRP Publication No. 8, 1965, p. 4,15.
70. ICRP Publication No. 14, 1969, p. 10.
71. ICRP Publication No. 14, 1969, p. 11.
72. ICRP Publication No. 14, 1969, p. 23.
73. ICRP Publication No. 14, 1969, p. 28.
74. ICRP Publication No. 14, 1969, p. 31, 32.
75. ICRP Publication No. 14, 1969, p. 37.
76. ICRP Publication No. 14, 1969, p. 57.
77. ICRP Publication No. 14, 1969, p. 112, 113.
78. ICRP Publication No. 14, 1969, p. 115, 116.
79. ICRP Publication No. 1973, p. 3.
80. ICRP Publication No. 22, 1973, p. 10, 11.
81. ICRP Publication No. 22, 1973, p. 12.
82. ICRP Publication No. 22, 1973, p. 13.
83. ICRP Publication No. 22, 1973, p. 14, 15.
84. ICRP Publication No. 26, 1977, p. 5.
85. ICRP Publication No. 26, 1977, p. 17, item 84.
86. ICRP Publication No. 26, 1977, p. 21.
87. ICRP Publication No. 26, 1977, p. 45–47.
88. ICRP Publication No. 26, 1977, p. 14, item 41.
89. ICRP Publication No. 39, 1984, p. 1.
90. ICRP Publication No. 39, 1984, p. 2.
91. ICRP Publication No. 39, 1984, p. 4.
92. ICRP Publication No. 39, 1984, p. 7.
93. ICRP Publication No. 39, 1984, p. IV.
94. ICRP Publication No. 39, 1984, p. I.
95. S. Jablon, "Letters." *Science,* 213, Sept. 20 1983, p. 6,7.
96. S. Jablon, "Die Grenzen der Strahlenbelastung." SVA Meeting of March 23, 1973, Zurich-Oerlikon, Switzerland.
97. M. Kaku and J. Trainer, *Nuclear Power, Both Sides,* Norton Company, New York, 1982, p. 109–133.
98. *Kettenreaktion,* No. 7. March 1984 "Ein Verein zur Untestützung der Kernenergie." Alpenstrasse 63, 3084 Wabern, Switzerland.
99. D. Kistner, *Radionuklide und Lebensmittel,* Bundesforschungsanstalt für Lebensmittelforschung, Karlsruhe, 1962.

100. "KÜR (Kommission zur Überwachung der Radioaktivität, Schweiz [Radiation Monitoring Commission of Switzerland])," Report 1973, p. 104.
101. KÜR Report 1983, p. 2, 3.
102. KÜR Report 1983, p. 8.
103. KÜR Report 1983, p. 12.
104. "State Legislature of Baden-Württemberg 8th Term," Drucksache 8/4482, Ministry of Labor, Health and Social Affairs, Nov. 22, 1983.
105. L. Lave et al., *Low-Level Radiation and US-Mortality.* "Working Paper No. 19-701," Carnegie Mellon University, Pittsburgh, PA, July 1971.
106. Levy et al., "Radiation-Induced F-Center and Coloidal Formation in Synthetic NaCl and Natural Rock Salt." *Nuclear Instruments and Methods,* B1.,198.
107. R. Lewis, "Shippingport the Killer Reactor?" *New Scientist,* Sep. 6 1973, p. 552, 553.
108. J.B. Little et al., "Plutonium-238 Exposure and Lung Cancer in Hamsters." *Science,* 138, 1975, p. 737.
109. K. Lorenz, "Über Gott und die Welt." *Natur,* No. 6, 1981, p. 27.
110. R.B. MacCandle et al., *Exp. Lung Cancer,* Springer Verlag, 1974, p. 485.
111. G. MacLeod. "TMI and the Politics of Public Health." Prepared for Presentation for the New York City and Chapters of Physicians for Social Responsibility on Nov. 22, 1980, Columbia University, International Affairs Auditorium, New York City.
112. B. MacMahon, "Prenatal X-ray Exposure and Childhood Cancers." *J. Nat. Cancer Inst.,* No. 28, p. 1173–1191.
113. T.F. Mancuso et al., "Radiation Exposure of Hanford Workers Dying from Cancer and Other Causes." *Health Physics,* 33, 1977, p. 369.
114. B. Manstein, *Im Würgegriff des Fortschritts,* Verlagsanstalt, 1961, p. 167.
115. B. Manstein, *Strahlen* S. Fischer Verlag, Frankfurt, 1977, p. 47, 49–51.
116. E. Marshall, "New A-Bomb Studies Alter Radiation Estimates." *Science,* 2121, May 22 1981, p. 900–903.
117. E. Marshall, "Japanese A-Bomb Data Will Be Revised," *Science,* 214, Oct. 2 1981, p. 31, 32.
118. C.W. Mays, *Proc. 3rd International Cong. IRPA,* Sept. 1973.
119. C. Mehring, "Immunitätslage der Bevölkerung nach Erhöhung der Umweltradioaktivität." Vitalstoffkongreß, Montreux, Sept. 12 1972; and C. Mehring, "Die biologischen Folgen der nuklearen Waffentests." *Protectio Vitae,* June 1972, pp. 220–225.
120. K.Z. Morgan, "Cancer and Low-Level Ionizing Radiation," *Bulletin of the Atomic Scientists,* Sept. 1978, p. 30–41.
121. Nagra, *Nagra aktuell,* No. 9, Nagra, 5401 Baden, Switzerland, Sept. 9, 1984.
122. Nagra, *Nagra aktuell,* No. 1, Nagra, 5491 Baden, Switzerland, Jan. 1985.

123. T. Najaran and T. Colton, "Mortality from Leukemia and Cancer." *The Lancet,* May 13 1978, p. 1018.

124. A. Nakaoka et al., "Evaluation of Radiation Dose from a Coal-Fired Power Plant." *Health Physics,* Vol. 48, Febr. 1985.

125. *Neue Revue,* "Atompfusch in Gorleben bedroht uns all." 4.1.1985.

126. *Neue Zürcher Zeitung,* "Die Kontroverse über Kernenergie in den USA." Oct. 30 1972.

127. *Neue Zürcher Zeitung,* "Aufregung um Rheumamittel." Jan. 7–8, 1984.

128. *Neue Zürcher Zeitung,* "Schadenersatz für Opfer von Atomtests." May 12–13 1984.

129. *Neue Zürcher Zeitung,* "Atomenergie-Investitionsruinen in Amerika." April 28–29, 1984.

130. *Neue Zürcher Zeitung,* "Eingeschrankter Verbrauch von Rheumamitteln. No. 79, Apr. 4 1985.

131. *New Scientist,* "Radiation Experts Row Over The Lethal Dose." Apr. 14, 1983.

132. *New Scientist,* "High Cancer Rates Found in Nuclear Plants." Oct. 11 984. 0. 3.

133. *New York Times,* Jan. 6 1984.

134. L. Pauling, "Genetic and Somatic Effects of Carbon-14." *Science,* Nov. 14 1958, p. 1183–1186.

135. A. Petkau, "Radiation Effects with a Model Lipid Membrance." *Canadian Journal of Chemistry,* Vol. 49 1971, p. 1187–1196.

136. A. Petkau, "Effect of Na^{22} on a Phospholipid Membrane." *Health Physics,* 22, 1972, p. 239–244.

137. R.O. Pohl, *Health Impact of Carbon-14,* Laboratory of Atomic and Solid State Physics, Cornell University, Ithaca, New York 14853.

138. G.G. Polykarpov, *Radioecology of Aquatic Organisms,* North Holland Publishing Company, Amsterdam, Reinhold Book Div., New York, 1966.

139. E. Radford, "New A-Bomb Data Shown to Radiation Experts." *Science,* 2212, June 19 1981, p. 1365.

140. L. Rausch, *Mensch und Strahlenwirkung,* Piper-Verlag, Frankfurt, 1977, p. 112/ 113.

141. "Report by the Governor's Fact Finding Committee." *Shippingport Nuclear Power Station,* Harrisburg, PA, 1974.

142. B. Rimland and G.E. Larson, "The Manpower Quality Decline." *Armed Forces and Society,* Fall 1981, p. 21–78.

143. A.E. Ringwood et al., "Stress Corrosion in a Borsilicate Glass Nuclear Wasteform." *Nature,* 311, 1984.

144. J. Rotblat, "The Risks of Atomic Workers." *Bulletin of the Atomic Scientists,* Sept. 1978, p. 46.

145. J. Rotblat, "Hazards of Low-Level Radiation." *Bulletin of the Atomic Scientists,* June–July 1981, p. 32–36.

146. M. Ruf, Bayrische Biologische Versuchsanstalt, Munich, "Die radioaktive Abfallbeseitigung aus Atomreaktoren in die menschliche Umwelt, mit besonderer Berücksichitgung der Gewässer." *Zentralblatt für Veterinärmedizin,* Beiheft 11, 1970.

147. M. Ruf, "Eliminierungs- und Rekonzentrierungsvorgänge bei der Ableitung von radioaktiven Abfallprodukten in Oberflächengewässer." Nabd 22, Verlag Oldenburg, Munich.

148. W.L. Russel, "Studies in Mammalian Radiation Genetics." *Nucleonics,* 23, 1965, p. 53–62.

149. C.I. Sanders, "Carcinogenicity of Inhaled Plutonium-238 in the Rat." *Radiation Research,* 56, 1973, p. 973.

150. *Science,* "Assessing the Effects of a Nuclear Accident." Vol. 228, April 5 1985, p. 31–33.

151. Scott et al., *Occupational X-Ray Exposure with increased Uptake of Rubidium by Cells.*

152. M. Segl and M. Kurihara, "Cancer Mortality for Selected Sites in 24 Countries." *Japan Cancer Society,* Tohoku University, Japan, Nov. 1972.

153. B. Shapiro and G. Kollmann, "Nature of Cell Membrane Injury to Irridated Human Erythrocytes." *Radiation Res. 34,* 1968, p. 335.

154. W. Sutow et al., "Growth Status of Children Exposed to Fallout Radiation on the Marshall Islands." *Pediatrics,* 1965, No. 36, p. 721–723.

155. R. Schleicher, "*Atomenergie, die große Pleite,* AVA-Buch 2000, Postfach 89, Affoltern a/A, Switzerland.

156. I. Schmitz-Feuerhake, *Die Wirkung ionisierender Strahlung auf den Mensch,* Part A, No. 8, 1979, University of Bremen.

157. G. Schwab, *Der Tanz mit dem Teufel,* Verein für Lebenskunde, Postfach 6, 5033 Salzburg, Austria, 1958.

158. Swiss Federal Council, "Schriftliche Beantwortung der Motion Schalcher, No. 76'391, 23.6.1976. Kernkraftwerke, Immissionen."

159. E.J. Sternglass, "Cancer: Relation of Prenatal Radiation to Development of the Disease in Childhood," *Science,* June 1963, p. 1100–1104.

160. E.J. Sternglass, "Infant Mortality and Nuclear Power Generation." "Hearings of the Pennsylvania State Committee on Reactor Sitting," Harrisburg, Oct. 1970.

161. E.J. Sternglass, "Infant Mortality Changes Near a Fuel Reprocessing Plant." "Testimony before the Illinois Pollution Control Board," Norris, Illinois, Dec. 10 1970.

162. E.J. Sternglass, *Proceedings of the Sixth Berkeley Symposium,* University of California Press, 1970.

163. E.J. Sternglass, *Infant Mortality Changes near the Big Rock Point Nuclear Reactor Power Station, Charlevoix.* Dep. of Radiology, University of Pittsburgh, PA. Jan. 6 1971.

164. E.J. Sternglass, *Infant Mortality Changes Near the Peach Bottom Nuclear Power Station in New York County.* Dep. of Radiology, University of Pittsburgh," Feb. 7 1971.

165. E.J. Sternglass, *Low-Level Radiation,* Ballantine Books, New York, 1972.

166. E.J. Sternglass, *Significance of Radiation Monitoring Results for the Shippingport Nuclear Reactor.* Dep. of Radiology, University of Pittsburgh, PA. Jan. 21 1973.

167. E.J. Sternglass, *Radioactive Waste Discharges from Shippingport Nuclear Power Station and Changes in Cancer Mortality.* Dep. of Radiology, University of Pittsburgh," PA. May 8 1973.

168. E.J. Sternglass, *Radioactive Waste Discharges from Shippingport Nuclear Power Station and Changes in Cancer Mortality.* Dep. of Radiology, University of Pittsburgh," PA. May 8 1973.

169. E.J. Sternglass, *Testimony Relating to Health Effects of Shippingport Nuclear Power Station.* Dep. of Radiology, University of Pittsburgh," PA. July 31 1973.

170. E.J. Sternglass, *Environmental Radiation and Cell Membrane Damage.* Dep. of Radiology, University of Pittsburgh, PA. Feb. 28 1974.

171. E.J. Sternglass, "Implications of Dose-Rate Dependent Cell-Membrane Damage for the Biological Effect of Medical and Environmental Radiation." "Proceedings of the Symp. on Population Exposure," Knoxville, Ten. Aug. 21 1974.

172. E.J. Sternglass, "Recent Evidence for Cell-Membrane Damage from Environmental Radiation." "Testimony at EPA Hearings on Radiation Standards for the Nuclear Cycle." Wash. DC, March 10, 1976.

173. E.J. Sternglass, "Health Effects of Environmental Radiation." Cincinnati Engineers and Scientists," Vol. 2, Oct. 1977.

174. E.J. Sternglass in J.O.M. Bockris, *Environmental Chemistry,* Plenum Press, New York and London, 1977, Chapter 15, p. 489.

175. E.J. Sternglass and S. Bell, "Fallout and the Decline of Scholastic Aptitude Scores." Presented at the Annual Meeting of the American Psychological Ass. New York, NY, Sep. 3. 1979.

176. E.J. Sternglass, "Infant Mortality Changes following the Three Mile Island Accident," Presented at the 5th World Concress of Engineers and Architects, Tel Aviv, Israel, Jan. 25, 1980.

177. E.J. Sternglass, *Secret Fallout,* McGraw-Hill Book Company, New York, 1981.

178. E.J. Sternglass, *Fallout and SAT Scores, Evidence for Cognitive Damage during Early Infancy,* Phi Delta Kappan, 64, April 1983, p. 539–545.

179. A.M. Stewart and G.W. Kneale, "Mortality Experiences of A-Bomb Survivors." *Bulletin of the Atomic Scientists,* May 1984, p. 62,63.

180. R. Stobaugh and D. Yergin, *Energie Report der Harvard Business School,* C. Bertelsmann Verlag, Munich, 1980.

181. T. Stokke, et al., "Effect of Small Doses of Radioactive Strontium on the Bone marrow." *Acta Radiologica,* 7, 1968, p. 321.

182. H. Strohm, *Friedlich in die Katastrophe* Verlag Association, Hamburg, 1973, pp. 30/31.

183. H. Strohm, *Friedlich in die Katastrophe* Verlag 2001, 6000 Frankfurt, W. Germany, 1981, p. 186.

184. *Tages Anzeiger,* "Ausverkauf von Kernkraftanlagen in Amerika." June 1 1984, Zürich.

185. *Tages Anzeiger,* "Schnelle Brüter unwirtschaftich." Zürich, Dec. 14, 1984.

186. D. Teufel, *Waldsterben, Natürliche und kerntechnisch erzeugte Radioaktivität,* "IFEU-Report" No. 25, 1983, Heidelberg, p. 32–32b.

187. *The Nuclear Engineer,* Vol. 24, No. 3, June 1983.

188. Tokanuga et al., *Breast Cancer in Japanese A-Bomb- survivors. Lancet,* 1982, p. 924.

189. L. Torrey "Radiation Cloud Over Nuclear Power." *New Scientist,* Apr. 24 1980, p. 197–199.

190. R. Tredici, *Die Menschen von Harrisburg,* Verlag 2001, Frankfurt, 1982.

191. S. Tschumi, *Symposium über den Schutz unseres Lebens,* Nov. 10-12 1970, ETH, Zürich.

192. S. Tschumi, *Allgemeine Biologie,* Verlag Sauerländer, Aarau, 1970.

193. J. Tseng, *Statistical Investigations of Possible Relationship between Nuclear Facilities and Infant Mortality.* Northwestern University, Evanston, Illinois, June, 1972.

194. UNSCEAR 1962, p. 10.

195. UNSCEAR 1962, p. 34.

196. UNSCEAR 1962, p. 145.

197. UNSCEAR 1962, p. 7.

198. UNSCEAR 1962, p. 122.

199. UNSCEAR 1972, Vol. II, p. 403, 404.

200. UNSCEAR 1977, p. 27.

201. UNSCEAR 1982, p. 8,9.

202. UNSCEAR 1982, p. 11.

203. UNSCEAR 1982, p. 16, item 88.

204. UNSCEAR 1982, p. 27.

205. UNSCEAR 1982, p. 30.

206. L. Von Middlesworth, "Small Quantities of I-131 in Thyroids of Sheep from Wales." *Health Physics,* Vol. 40, April 1981, p. 525.

207. M. Wahlen et al., "Radioactive Plume from the TMI Accident, Xenon-133 in Air at a Distance of 375 Kilometers." *Science,* Feb. 8 1980, p. 639.

208. S. Weish and E. Gruber, *Radioaktivät und Umwelt,* 2nd Edit., G. Fischer Verlag, Stuttgart, W. Germany 1979.

209. M. Wenz, "Gorleben versalzen." *Natur,* No. 3, Verlag Ringier, Zofingen, Switzerland, March 1985.

210. J.P. Wesely, "Background Radiation as the Cause of Fatal Congenital Malformation." *Intern. J. Rad. Bull.,* 2, 1960, p. 297.

211. *Zofinger Tagblatt,* "Britische Soldaten als Opfer." Jan. 10 1983.

212. *Zofinger Tagblatt,* "'Gewähr' im Wandel der Zeit." June 25 1984.

213. *Zofinger Tagblatt,* "Mexiko revidierte Energieplanung." Aug. 16 1984.

214. [non-existant]

215. K.Z. Morgan, "ICRP Risk Estimates—An Alternative View." WISE-Amsterdam, PO Box 5627, 1007AP Amsterdam, The Netherlands, Nov. 24 1986.

217. *Tagesanzeiger,* "Risiken genetischer Schaden ist gering." Zurich, May 4 1987.

218. Jay M. Gould, *U.S. Mortality and Three-Mile Island.* Public Data Access, Inc., 30 Irving Pl., New York, 10003, released May 24 1987.

219. A. Petkau, "Radiation Carcinogenesis from a membrane perspective." *Acta Physical. Scand.,* 1980, 492, 81-90.

220. A. Petkau, *Protection and Repair of Irradiated Membranes: Free Radicals, Aging and Degenerative Diseases,* Alan R. Liess Inc., 1986, pp. 481–508.

221. Carl J. Johnson, "Some Studies of Low-Level Radiation and Cancer in the United States." Presented at the Universities of Basle, Zurich and Lausanne; contains 47 references, June 1987.

222. E.J. Sternglass and J. Scheer "Radiation Exposure of Bone Marrow Cells to Strontium-90 during early development as a possible Cofactor in the Etiology of AIDS." paper delivered at the Annual Meeting of the American Association for the Advancement of Science, Philadelphia, May 29 1986.

223. E.J. Sternglass "The Implications of Chernobyl for Human Health," *International Journal of Biosocial Research,* v. 8 # 1, July 1986, p. 7-36.

224. O. Haller and H. Wigzell, "Suppression of NK Cell Activity with Radioactive Strontium: Effector Cells are Marrow Dependent." *J. Immunology,* 1977, p. 118.

225. Jean L. Merx, "Oxygen Free Radicals Linked to Many Diseases." *Science,* vol. 235, Jan. 30 1987, p. 529–531.

226. A.D. Sakharov, "Radioactive Carbon from Nuclear Explosions and Non-threshold Biological Effects." *The Soviet Journal of Atomic Energy,* 4, p. 757–762, 1958.

227. O. Zaguri, J. Bernard, R. Leonard et al., "Long-Term Cultures of HTLV-III-Infected Cells: A Model Cytopathology of T-Cell Depletion in AIDS. *Science* 231, 1986, p. 850–853.

Part III and IV

1. W.D. Barth, "Zwischen den Waldschäden im Odenwald und dem KKW Obrigheim gibt es keinen Zusammhang." *Basler Zeitung.* 12.4. 1984. Manuscript with the same Title obtainable from: Kernkraftwerk Obrigheim, Postfach 100, 6951 Obrigheim, W. Germany, March 1984.
2. *Basler Zeitung,* "Der japanische Wald hat keine Lobby." 10.1. 1985.
3. BEIR 1972, p. 22, 30.
4. Bild, 1984, 10, 4.
5. C.H. Binswanger "Umweltkrise und National-Okonomie." *Neue Zürcher Zeitung,* June 4 1972.
6. W.L. Boeck, "Meterological Consequences of Atmospheric Krypton 85." *Science,* 193. July 16, 1976, p. 195-197.
7. H. Bonka, *Strahlenbelastung der Beyölkerung durch Emissionen aus Kernkraftwerken im Normalbetrieb,* Verlag TUV Rheinland 1982.
8. C. Bosch, *Die sterbenden Wälder,* C.H. Becksche Verlagsbuchhaltung, Munich, 1983, p. 21, 22.
9. J.B. Bucher, "Bemerkungen zum Waldsterben und Umweltschutz in der Schweiz." *Forstwissenschaftliches Centralblatt,* April 1984, p. 23, 24.
10. Buchmann, *Chem. Communications,* 1970, p. 1631.
11. *Bulletin of the Atomic Scientists,* "Tritium Warning," March 1984.
12. Federal Environmental Protection Agency, *Luftbelastung 1983.* EDMS, Bern, Switzerland, Sept. 1982.
13. Federal Ministry of Food, Agriculture and Forestry, *Waldsterben durch Luftverunreinigung.* Pamphlet 273, Landwirtschaftsverlag, Monchen-Hiltrug, West Germany, 1982.
14. Federal Ministry of the Interior, *Waldschäden und Luftverunreinigungen.* Special Report, March 1983, Rat für Umweltfragen, Verlag Kohlhammer, Stuttgart, July 1983, p. 73, 71, 84, 19, 30.
15. W. Burkhart, "Waldsterben, auch hier der schwarze Peter bei den A-Werken¿" *Basler Zeitung,* Nov. 30 1983.
16. W. Burkhart, "Nochmals Radioaktivität und Wald." *Basler Zeitung,* Feb. 17, 1984.
17. Commission of the European Communities, *Acid Rain.* Graham and Trotman, Ltd., London SWIV, IDE. 1983.
18. Commission of the European Communities, *European Seminar on the Risks from Tritium Exposure,* Mol, Belgium, 22.–24. Report EUR 9065 en. 1984, Office of the European Communities, Luxemburg, Nov. 1982.
19. U. Dekar and H. Thomas, "Unberechenbares Spiel der Natur: Die Chaos-Theorie." *Bild der Wissenschaft,* Deutsche Verlagsanstalt, Stuttgart, # 1 1983, p. 63-75.

20. Swiss Federal Government, "Auch der Schwedische Wald beginnt zu kränkeln." Bern, Feb. 4, 1985.
21. West German Bundestag, 10th Term, Drucksache 10/1730 of July 9 1984, Sachgebiet 2129.
22. EAWAG-Report 1984, "Regen und Nebel als Trager umweltbeeinträchtigender Stoffe." Swiss Federal Water Supply, Sewage-Treatment and Water-Protection Institute (EAWAG), 8600 Dübendorf, Switzerland.
23. Swiss Federal Department of the Interior, "Waldsterben und Luftverschmutzung." Bern, 1984.
24. Swiss Federal Institute for Forestry Research (EAFV), "Waldschäden in der Schweiz." Oct. 1984, p. 817–831.
25. *Encyklopädie Naturwissenschaft und Technik,* Verlag moderne Industrie, Wolfgang & Co., 5912 Landsberg a. L., 1981, p. 4636–4637.
26. A.W. Fairhall, "Potential Impact of Radiocarbon Release by the Nuclear Power Industry." Washington University, Seattle, 1980.
27. F.N. Flakus, "Symposium of Problems of Radiation Protection in Connection with the use of Tritium and Carbon-14, and their Compounds." Berlin, IAEA, Box 200, 1400 Vienna, Austria, Nov. 14–16, 1979.
28. J. Fuhrer, "Atmosphärische Einflußfaktoren der Waldschädigung." Informational Meeting on Forest Damage due to Emissions, GDI Institute, 8803 Ruschlikon, Switzerland, Nov. 29 1982.
29. J. Fuhrer, "Formation of Secondary Air Pollutants and their Occurrence in Europe." *Experientia* 41, Birkhauser-Verlag, 4010 Basel, Switzerland, 1985, p. 286–301.
30. F. Funk and S. Person, *Science* # 166, 1969, p. 1629.
31. O. J. Fuhrer, "Maßnahmen und Verordnungen—Erste Schritte zur Neuorientierung." *GDI-Schriften # 35* "Stirbt der Boden¿" GDI. Inst., 8803 Rüschlikon, Switzerland, Jan. 19-20, 1984.
32. H.W. Georgii, *Global distribution of the Acidity in Precipitation, Deposition of Atmospheric Pollutants.* Dortracht, 1982.
33. R. Graeub, "Atomenergie, Mitverursacher des Waldsterbens¿" *Basler Zeitung,* Nov. 5, 1983.
34. R. Graeub, "Waldsterben und Radioaktivitat." *Basler Zeitung,* Jan. 19 1984.

35. R. Graeub, "Waldschäden durch Atomanlagen¿" *Basler Zeitung,* Mar. 9 1984.
36. W. R. Guild, "Hazards from Isotopic Tracers." *Science,* 128, p. 1308, 1958.
37. J.E. Hall et al., "The Relative Biological Effectiveness of Tritium Beta Particles Compared to Gamma Radiation—Its Dependence on Dose-Rate." *Brit. Journal of Radiology,* 40 1967, p. 704–710.
38. G.R. Hendrey, "Automobiles and Acid Rain." Terrestrial Land Aquatic Ecology Div. Brookhaven National Laboratory, Upton, L.I., New York, *Science,* 222, Oct. 7 1983.

39. A. Hofmann, "Pflanzenkundliche Überlegungen zum Waldsterben." *Basler Zeitung,* Oct. 6 1983.

40. E. Hollstein, "Säkulärvariationen des Eichenwuchses in Mitteleuropa." Kolloquium in Trier, University of Trier, May 15–17, 1980.

41. H. Hommel and Kas, G., "Elektromagnetische Verträglichkeit des Biosystems Pflanze." *Allg. Forst-Zeitung,* No. 8, 1985, p. 172–174.

42. J.W. Horbeck, "Acid-Rain & Facts and Fallacies." *Journ. of Forestry,* 79, 1981, p. 438–443.

43. A. Hüttermann, "The Effects of Acid Deposition on the Physiology of the Forest Ecosystem." *Experientia,* 41, Birkauser Verlag, 4010 Basel, Switzerland, May 1985, p. 578, 583.

44. IAEA, "Tritium in Some Typical Ecosystems." *Technical Report No. 207,* Vienna 1981, p. 1–116.

45. S. Ichikawa, Presentation, Apr. 1 1977 in Salzburg, Bürgerinitiative Lübeck e.V., Postfach 1926, 1400 Lubeck, Austria.

46. *Information des Zentralverbandes,* Burgerinitiativen gegen Atomgefahren. No. 1/84 (signed by Prof. Dr. H. Noller, Univ. Prof. of Physical Chemistry, Vienna), Verlagspostamt 1128, Vienna, Austria.

47. A. H. Johnson et al., "Acid Deposition and Forst Decline." *Envir. Sci. Technol.,* 17, 1983, p. 294–305.

48. H.S. Johnston, "Human effects on the global atmosphere." *Ann. Rev. Phys. Hem.,* 35, p. 481–505.

49. O. Kandler, "Waldsterben, Emissions- oder Epidemie-Hypothese¿" *Naturwissenschaftliche Rundschau,* Nov. 1983, p. 488–490.

50. Kirchmann et al., *Health Physics,* 21, 1971, p. 61–66.

51. R. Kirchmann, M. Molls, C. Streiffer, and J. Mewissen, *Colloquium on the Toxicity of Radionuclides,* Liege, Societe belge de Radiobiologie, Nov. 19–20, 1982.

52. R. Kollert, *Kerntechnik und Waldschäden,* Study under contract to the Stiftung mittlere Technologie, Kaiserslautern, Bremen/Perzelle, 1985.

53. L.A. König, et al. *Kerntechnik und Waldschäden,* Karlsruhe Nuclear Research Center, KFK 3704, March 1984.

54. L.A. König, "Umweltradioaktivat und Kerntechnik als mögliche Ursachen von Waldschäden¿" *KKF-Nachr.,* vol. 17, 1/1985, p. 22–31.

55. KÜR, "25 Jahre Radioaktivitatsüberwachung in der Schweiz." KUR, c/o Physics Inst. of the Perolles, University, Freiburg, Switzerland, Nov. 1982.

56. KÜR, Annual Report 1982.

57. KÜR, Annual Report 1983.

58. S. Laughlin and O.U. Braker, "Methods for Evaluating and Predicting Forest Growth Responses to Air Pollution." *Experientia,* Vol. 41. Mar. 15, 1985, p. 310–319.

59. P. Leuthol, "ETH-Forschungsprojekt Manto: *Drahtlose Nachrichtenübertragung eine Gefahr für die Umwelt¿* Interim Report 2, Study 2, Institut for Kommunikationstechnik, ETHZ, Zurich, Switzerland, Dec. 24, 1984.

60. Levin et al., "The Effect of Anthropogenic CO_2 and C-14 Sources on the Distribution of C-14 in the Atmosphere." *Radiocarbon,* Feb. 22, 1980, p. 379–391.

61. H. Loosli, "Haben künstlich erzeugte Radionuklide wie Kr-85, C-14, H-3 mit der Luftionisation, mit dem sauren Regen und dem Waldsterben zu tun¿" *SVA-Bulletin,* No. 3/84.

62. B.J. Mason et al., *Environmental contamination by Radioactive Materials,* IAEA, Vienna, 1969.

63. A. Meier and M. Wallenschus, "Tradescantia, ein Bioindikator für Radioaktivitat." University of Bremen, *Information zu Energie u. Umwelt,* Part A, No. 18.

64. I.W. Mericle et al., "Cumulative Radiation Damage in Oak Trees," *Radiation Botany,* No. 2. 1962, p. 265–271.

65. H. Messerschmidt, "Anmerkungen, Fragen und Kritik zu der Veröffentlichung KFZ 3704, March 1984, Kerntechnik und Waldschaden des Kernforschungszentrums Karlsruhe GmbH." Manuscript by H. Messerschmidt, 3130 Luchow, W. Germany, Sept. 1, 1984/Nov. 20 1984.

66. H. Metzner, *Waldsterben durch Kerntechnische Anlagen¿* Literature Review for the Ministry of Food, Agriculture, Environment and Forestry of the State of Baden-Wurttemberg, Institut fur Chemische Pflanzenphysiologie, University of Tubingen, Sept. 1985.

67. *Natur,* "Auch Atomkraft schuldig¿," # 3, 1984.

68. *Natur,* "Auch Atomkraft schuldig¿," # 11 1983 and # 1, 1984.

69. *Natur,* "Mit Strahlung geht's schneller." No. 8, 1984.

70. *Neue Zürcher Zeitung.* "Japan ohne Waldschaden¿" Forschung und Technik, No. 278, Nov. 1984.

71. *Neue Zürcher Zeitung.* "Waldsterben Worte oder Taten¿," Jan. 18 1985.

72. *Neue Zürcher Zeitung.* "Das Atomgewerbe zum Waldsterben." Feb. 4, 1985.

73. *Neue Zürcher Zeitung.* "Sind Radiowellen für Pflanzen schädlich¿: Apr. 20 1985.

74. *New York Times,* "Widespread Ills Found in Forests in Eastern U.S." Feb. 2 1984.

75. M. Oehen, "Tritium Umweltbelastrung." Inquiry No. 83.952, Response of the Federal Council of Switzerland, Dec. 15 1983.

76. R.L. Otlet, "The Use of C-14 in Natural Materials to Establish the Average Gaseous Dispersion Patterns of Releases from Nuclear Installations." *Radiocarbon 25.2.1983,* p. 592–602.

77. R. Pohl, "Health Impact of Carbon-14." *Nuclear Energy,* Laboratory of Atomic and Solid State Physics, Cornell University, Ithaca, New York 14853, 1975.

78. G. Reichelt, "Der sterbende Wald in Süddeutschland und Ostfrankreich." *BUND-Information,* No. 25, 1983, Stuttgart.

79. G. Reichelt, "Zur Frage des Waldsterbens in Frankreich." *Landschaft und Stadt,* No. 4, 1983, Verlag Dugen Ulmer, Stuttgart, p. 150, 162.

80. G. Reichelt, "Der sterbende Wald in Süddeutschland und Ostfankreich." *BUND-Information,* No. 25, Stuttgart, 1983.

81. G. Reichelt, "Modellrechnungen sollten sich eigentlich nach der Wirklichkeit richten." *Basler Zeitung,* April 12 1984.

82. G. Reichelt, "Zusammenhang zwischen Radioaktivitat und Waldsterben." Presentation, University of Hannover, May 25, 1984.

83. G. Reichelt, "Manuscript on Mapping done in May 1984, at the Nuclear Power Plants at Beznau Switzerland, Wurgassen, West Germany.

84. G. Reichelt, "Wo das Waldsterben begann." *Basler Zeitung,* Aug. 24 1984.

85. G. Reichelt, "Zur Frage des Zusammenhangs zwischen Waldschäden und dem Betrieb von Atomanlagen vorlaüfige Mitteilung." *Forstwissenschaftliches Centralblatt,* Sept. 1984, p. 290–297.

86. G. Reichelt, "Waldschadensmuster im Umkreis uranerzhaltiger Gruben und ihre Interpretation." *Allg. Forst-und Jagdzeitung,* No. 7/8, 1984, p. 184–190.

87. G. Reichelt, "Zusammenhänge zwischen Radioaktivitat und Waldsterben¿" *Okologische Konzepte,* No. 20, 1984, George Michael Pfaff Gedächtnisstiftung, 6750 Kaiserslautern, West Germany.

88. G. Reichelt, "Waldschadensmuster im Umkreis atomtechnischer und industrieller Anlagen im Vergleich zu industrieferneren Gebieten." Study Contract, Aug. 3 1984 of the Ministry for Food, Agriculture, Environment and Forests, Baden Wurttemberg. Manuscript obtained June 13 1985 by Prof. Reichelt, Uhlandstr. 35, 7710 Donau-Eschingen, W. Germany.

89. Reiter et al., *3rd Eur. Symposium on Physico-Chemical Behaviour of Atmospheric Pollutants,* Eds. Versino and Angletti D. Reidel Publ. Comp. Dordrecht, the Netherlands, Varese 10-12, April 1984, p. 480–481.

90. Reitz and Kopp, *Zeitschrift für physikal. Chemie.* A 179 (1937), 126, 184 (1939), 430.

91. K.J. Seelig, *12-Punkte-Programm,* Lindau 7.9.6.1983. Adresse, Dr. med. K.J. Seelig, Kornmarkt, 5521 Biersdorf, W. Germany.

92. K.J. Seelig, "Biopathogene, bislang verschwiegene, unbekannte Einflüsse des nuklearen Brennstoffzyklus auf die derzeitige Ökomisere," Nov. 1983.

93. K.J. Seelig, "Material an die verantwortlichen Befürworter für großtechnische Nutzung der Kernenergie insbes. zu Deuterium, Tritium, C-14 u.a. gasförmigen Freisetzungen." Manuscript, Sept. 1984.

94. K.J. Seelig, "Waldsterben und radioaktive Abgase aus KKW." Manuscript, Nov. 28 1984.

95. M. Segl et al., "Anthropogenic C14-variations." *Radiocarbon,* Feb. 25 1983, p. 583–592.

96. C. Seigneur et al. "Computer Simulation of the Atmospheric Chemistry of Sulfate and Nitrate Formation." *Science,* 225, 1984, p. 1028–1029.

97. Shell Switzerland, "Mitteilung, August 1984." Badenerstr. 66, Zurich.

98. S. Soom, "Memorandum Soom, mit Materialien, Fragen und Meinungen zum Thema Waldsterben und Radioaktivitat." Available from S. Soom, Ackerstr. 8, Nussbaumen 1983/1984), Switzerland.

99. A. H. Sparrow, "Tolerance of Certain Higher Plants to Chronic Exposure to Gamma Radiation from Cobalt." *Science,* No. 118, Dec. 4 1953, p. 698–698.

100. A. H. Sparrow et al., "The Effects of External Gamma Radiation from Radioactive Fallout on Plants with special References to Production." *Radiation Botany,* Vol. 11, 1971, p. 85–118.

101. *Spiegel,* "Le Waldsterben." Oct. 15 1984, p. 186.

102. Subba Ramu, M.C., *Ethylene in the Atmosphere and its Role in Aerosol Formation,* Bhabba Atomic Research Centre, Bombay, India, No. 1128, 1981.

103. H.E. Suss, "Ist die Sonnenaktivität für Klimaschwankungen verantwortlich¿" *Um-schau,* 1979, p. 312–316.

104. Symposium *Electronic Compatibility.* 5.-7. March 1985, ETH-Zentrum, 8092 Zurich, Bertaud A.J., p. 213–216; also Chen Q. et al. p. 199–204.

105. Schmitz et al., *Emission von Radionukliden aus den Halden des alten Silber-Kobalt-Erzbergbaus von Wittichen.* Glückauf-Forschungshefte 43, 4, 1982, p. 145–154.

106. W. Schöpfer and J. Hradetzky, "Der Indizienbeweis, Luftverschmutzung massgebliche Ursache der Walderkrankung." *Forstwissenschaftliches Centralblatt,* Sept. 1984, p. 244.

107. E. Schuhmacher, *Small is Beautiful,* Blond and Brigg Ltd., London, 1974.

108. E. Schuhmacher, *Es geht auch anders,* Verlag Dash, Munich, 1973.

109. S. Schütt, *So stirbt der Wald,* BlV-Verlagsgenossenschaft, Munich, 1983.

110. S. Schütt, *Der Wald stirbt an Streß,* C. Bertelsmann Verlag, 1984.

111. H. Schüttelkopf, "Verhalten langlebiger Radionuklide in der Biosphäre." "Working Session on Radioecology of the German Atomic Forum," Oct. 2-3, 1979.

112. G. Schwarz et al., *Possible Future Effects on the Population of the Federal Republic of Germany of Gaseous Radioactive Effluents from Nuclear Facilities,* IAEA, Vienna, 1975, p. 194–207.

113. F.H. Schwarzenbach, *Das Waldsterben als politische Herausforderung,* EAFV, 8903 Birmensdorf, Switzerland Aug. 3 1983.

114. F.H. Schweingruber, *Dichteschwankungen in Jahrringen von Nadelhölzern in Beziehung zu klimatisch-ökologischen Faktoren, oder das Problem der falschen Jahrringe.* Report 213, EAFV, 8903 Birmensdorf, Switzerland, May 1980.

115. F.H. Schweingruber, *Der Jahrring,* Verlag Haupt, Bern. 1983, p. 202, 204, 210-211.

116. F.H. Schweingruber, *Eine jahrringanalytische Studie zum Nadelbaumsterben in der Schweiz,* Report No. 253, EAFV, 8903 Birmensdorf, Switzerland, August 1983.

117. Swiss Society for Population Problems, Bern, *Rundschreiben,* Nov. 23 1973.

118. E.J. Sternglass, "Nuclear Power May Be Dangerous to Our Trees," *The New York Times,* March 13, 1983.

119. Stewart et al., *Tritium in Pine Trees from Selected Locations in the USA, Including Areas of Nuclear Facilities,* U.S. Geological Survey, Prof. Paper, 800-B, 1972, p. 265–271.

120. M. Stuiver, "Atmospheric C-14 Changes Resulting from Fossil Fuel CO_2 Releases and Cosmic Ray Flux Variability," *Earth and Planetary Sciences Letters,* 53, 1981, p. 348–382.

121. W. Stumm et al., "Der Nebel als Träger konzentrierter Schadstoffe." *Neue Zürcher Zeitung,* Jan. 16, 1985.

122. D. Teufel, *Waldsterben, natürliche und kerntechnisch, erzeugte Radioaktivitat,* IFEU-Report No. 25, IFEU-Inst. 6900 Heidelberg, W. Germany, 1983.

123. J.R. Trotter, "Hazard to Man of Carbon-14." *Science,* 128, Dec. 12 1958, p. 1490–1495.

124. I. Tripet and S.Wiederkehr, *Etude du problème des précipitations acides en Suisse,* Ecole Federale Polytechnique Lausanne, Inst. du Genie de l'Environement, March 1983.

125. S. Tschumi, *Allgemeine Biologie,* Verlag Sauerlander, 1970.

126. S. Tschumi, "Ursachen und Bekämpfung der Umweltkrise," *Techn. Rundschau,* 6.3.1974. Hallwag Verlag, Bern.

127. Ulmann, *Enzyclopädie der techn. Chemie,* 3rd Edit. vol. 2/1, Munich-Berlin, 1961, p. 955.

128. Ulmann, *Enzyclopädie der techn. Chemie,* 4th edit. vol. 6, Verlag Chemie, 6940 Weinheim, W. Germany, 1981, p. 226.

129. Federal Environmental Office, *Luftqualitätskriterien für photochemische Oxidantien.* Report 5, 1985, Berlin.

130. UNSCEAR 1982, p. 10.

131. UNSCEAR 1964, p. 13.

132. UNSCEAR 1969, No. 13, p. 19.

133. M. Urban, "Waldsterben auch durch Radioaktivitat?" *Süddeutsche Zeitung,* Munich, Apr. 25 1985.

134. K.G. Vohra, "Combined Effects of Radioactive Chemical and Thermal Releases on the Environment." Symposium held in Stockholm 2.-5. June, 1975, IAEA, Vienna 1975, p. 209-221.

135. A. Von Rotz, "Radioaktive Umweltverschmutzung und Vergiftung." *Mitgliederzeitung der Schweiz. Krankenkasse Helvetia,* Habegger AG, Solothurn No. 11, 1971.

136. Weish and Gruber, *Radioaktivität und Umwelt,* Gustav Fischer Verlag, Stuttgart, 1975.

137. W. Weiss et al. *Evidences of Pulsed Discharges of Tritium from Nuclear Energy Installations in Central European Precipitation.* Inst. for Environmental Physics of the University of Heidelberg. IAEA-SM 232/18, 1979.

138. A. Weiss, "Manuscript," for Prof. Reichelt (cf. Lit. Ziff.87), University of Munich, Apr. 4 1984.

139. M. Wenzel and S. Schulte, *Tritium-Markierungen nach der Wilzbach-Methode.* Walter de Gruyther-Verlag, Berlin, 1972.

140. F.W. Whicker and V. Schultz, *Radioecology, Nuclear Energy and the Environment,* Vol LL. CRC-Press Inc. Boca Raton, Florida, USA, 1982, p. 128, 153–162.

141. J. Wilzbach, *Amer. Chem. Soc., 79,* 1957, p. 1013.

142. WWF-Schweiz, *Schadkartierung an Fichten in der Umgebung der schweiz, Kernkraftwerke,* Büro für Forstwirtschaft und Umweltplanung, 8964 Rudolfstetten, Switzerland, June 1984.

143. J.Zavitovski, (Editor), *The Enterprise, Wisconsin Radiation Forest Radioecological Studies,* Inst. of Forest Genetics, North Central Service, U.S. Dep. of Agriculture, Rhinlander, Wisconsin, TID-26113-P2. 1977, p. iii, 141–165.

144. *Zofinger Tagblatt,* "Elsässische Gemeinden machen gegen das Waldsterben mobil." Zofingen, May 23 1985.

145. Federal Environmental Protection Agency, *Radioaktivität und Waldsterben,* Schriftenreihe Umweltschutz, No. 43, Bern, 1985.

146. Federal Environmental Protection Agency, *Radio- und Mikrowellen als mögliche Ursachen für Waldschäden,* Schriftenreihe Umweltschutz, No. 44, Bern, 1985.

147. Hauser, Bert, "Brisante Tübinger Studie." *Frankfurter Allg. Zeitung,* Jul 20 1985.

148. Federal Ministery of the Interior, *Umweltprobleme der Landwirtschaft,* March 1985, Rat für Umweltfragen, Verlag Kohlhammer, Stuttgart.

149. *Neue Zürcher Zeitung,* "Kernkraftwerke unschuldig am Waldsterben." Reply by the Federal Department of the Interior to Pamphlet No. 43, BUS-Schriftreihe "Umweltschutz." Aug. 24–25, 1985.

Part V

1. Robert J. Pellegrini: "Nuclear Fallout and Criminal Violence". *Int. J. of Biosocial Research,* 1987, 9,2: pp. 125–43.

2. Robert J. Pellegrini: Early-Development Exposure to Nuclear Fallout." *Psychological Reports,* 1988, 63, pp. 857–8.

3. J. H. Schroeder, Institut für Strahlenbiologie GFS, München-Neuherberg, cited in *Bild-Zeitung,* September 12, 1988, "Umweltgifte machen dumm."

4. Lengfelder E. "Strahlenwirkung, Strahlenrisiko" H. Hugendobler-Verlag (1988), Münich.

5. "Radioactivitätsmessungen in der Schweiz nach Tschernobyl und ihre wissenschaftlische Interpretation." University of Bern. 20-22, October 1986, *Tagungsbericht.* 1.1., pp. 739-740/1.2. pp. 553–60/1.3, p. 665/1.4 p. 319.

6. *BEIR* V: "Health Effects of Exposure to Low Levels of Ionizing Radiation." National Academic Press, pp. 17–20, 29–31, 361.

7. op cit. p. 20

8. op cit. p. 46

9. op cit. p.

10. *The Lancet:* "HIV infection in a Norwegian Family before 1970." June 11, 1988, p. 1344.

11. *The New York Times:* "Origin of Human Aids-Viruses May Be as Recent as 40 years ago." June 9,1988. Report by Temple F. Smith et al., Dana-Farber Cancer Institute, Boston.

12. Ernest J. Sternglass, Jens Scheer: "Radiation exposure of bone marrow cells to Strontium 90 during early development as a possible Cofactor in the Etiology of Aids." Paper delivered at the 1986 Meeting of the American Association for the Advancement of Science, Philadelphia, May 29, 1986.

13. Ernest J. Sternglass: "The Implication of Chernobyl for Human Health," *International Journal of Biosocial Research,* July 1986, Vol. 8, 1, pp. 7–36.

14. T. Stokke et al.: "Effect of Small Doses of Radioactive Strontium on the Bone Marrow" *Acta Radiologica,* 7, 1968, p. 321.

15. O. Haller and H. Wiegzell: "Suppression of Natural Killer Cell Activity with Radioactive Strontium: Effector Cells are Marrow-Dependent". J. *Immunol.,* 1977, p. 118.

16. R. Jensen: *Umweltschaden Aids¿,* 1990, Buch 2000, Affoltern a/Albis, Switzerland.

17. O. Zaguri, J. Bernard, R. Leonard et al.: "Long Term Cultures of HTLV-III Infected Cells: A Model Cytopathology of T-Cell Depletion in Aids." *Science* 231, 1986, pp. 853–59.

18. "Wochenend": "In der Hölle des Aids". No. 10–15, 1988, Hamburg.

19. Ernest J. Sternglass: "The Implication of Chernobyl for Human Health." *Int. Journal of Biophysical Research,* Vol 8, 1, pp. 7–36 (July, 1986).

20. Andrei D. Sakharov: "Radioactive carbon from Nuclear Explosions and Non-threshold Biological Effects," *The Soviet Journal of Atomic Energy*, 4, 1958, pp. 757–762.

21. IFEU-Bericht No. 52 1989. IFEU-Institut, Heidelberg.

22. Otto Hug Strahleninstitut, Im Rheingarten 7, Bonn D-530, Report No. 1, 1988.

23. E.P. Radford: "Recent Evidence of Radiation-induced Cancer in Japanese Atomic Bomb Survivors" in *Radiation and Health,* R. Jones and R. Southwood, eds., 1987, John Wiley & Sons Ltd., Chichester, pp. 87–96.

24. Karl Z. Morgan: "ICRP Risk Estimates: An Alternative View," in *Radiation and Health,* R. Jones and R. Southwood, eds., 1987, John Wiley & Sons Ltd., Chichester.

25. *Tages-Anzeiger* (Zürich), January 30 1991. "Grenzwerte im Strahlenschutz sind viel zu hoch".

26. ICRP, Draft February 1990 and ICRP 60 (1990).

27. W. Köhnlein, R.H. Nussbaum: "Das Krebsrisiko ist 10 mal grösser." Strehlentelex (No. 90/91 October 4 1990), Berlin.

28. W. Köhnlein and R.H. Nussbaum: "Reassessment of Radiogenic Cancer Risk and Mutagenesis at Low Doses of Ionizing Radiation", *Advances in Mutagenesis Research,* G. Obe editor. Vol. 3, 1991.

29. W. Köhnlein, H. Kuni. and Inge Schmitz–Feuerhake, editors: *Niedrigdosisstrahlung und Gesundheit.* (1990) Springer Verlag, Berlin, pp. 201–214, pp. 183–192.

30. W. Köhnlein and R. Nussbaum: "Die neue Krebsstatistik der Hiroshima-Nagasaki-Ueberlebenden". *Medizinische Klinik.* Band 86, Heft Nr. 2, 1991, February 15.

31. M.J. Gardner et al.: *British Medical Journal,* February 17 1990, Vol 300, pp. 423–32. P. Weish and E. Gruber: "Radioactivität und Umweld." Gustav Fisher Taschenbuch-Verlag, 3rd Ed., 1986.

33. Charles Waldren et al.: "Measurement of low levels of X-ray mutagenesis in relation to human disease" *Proc. Natl. Acad. Sci.,* 83, July, 1986, pp. 4839–43.

34. J.L. Marx: "Oxygen Free Radicals Linked to Many Diseases." *Science* January 30, 1987, pp. 529–531.

35. Ralph Graeub: "Die Realität des Petkau-Effekts". Lecture held at the Symposium "Die Wirkung niedriger Strahlendosen auf den Menschen" August 26–27, 1988, at the Institute for Radiation Biology, University of Münster (Germany). Reprints available by request from the author. Postfach 210, CH-8135, Langnau a. Albis, Switzerland.

36. W.S. Chelack, M.P. Forsyth, Abram Petkau : "Radiobiological properties of Acholeplasma laidlawii B". *Can. J. Microbiol.* 20, 307–320, 1974.

37. Abram Petkau, W.S. Chelack: "Radioprotective Effects of Cysteine" *Int. J. Radiobiol.,* 25, 321, 1975.

38. Abram Petkau, W.S. Chelack, S.D. Pleskach, T.P. Copps: "Radioprotection of Hematopoeitic and Mature Blood Cells by Superoxide Dismutase". Paper presented at the annual meeting of the Biophysical Society, Philadelphia, 1975.

39. Abram Petkau, W.S. Chelack, S.D. Pleskach, B.E. Meeker and L.M. Brady: "Radioprotection of mice by superoxide dismutase." Biochem. *Biophys. Res. Commun.* 1975, 65, pp. 88–93.

40. Abram Petkau, W.S. Chelack, S.D. Pleskach, C. Barefoot, and B.E. Meeker: "Radioprotection of bone-marrow stem cells by superoxide dismutase." *Biochem. Biophys. Res. Commun.* 1975, 67, pp. 1167–74.

41. Abram Petkau, K. Kelly, W.S. Chelack, S.D. Pleskach, C. Barefoot, and B.E. Meeker: "Radioprotective effect of superoxide dismutase on model phospholipid membranes." Biochem. Biophys. *Acta* 1975, 433, pp. 445–56.

42. Abram Petkau: "Radiation carcinogenesis from a membrane perspective," *Acta Physiol.* Scand. 1980. Suppl. 492, 81–90.

43. J.C. Szekely, K.A. Perry, Abram Petkau: "Simulated responses to log-normally distributed continuous low radiation doses," *Health Physics,* 45, 3, 1983, pp. 699–711.

44. Abram Petkau, C.A. Chuaqui: "Superoxide Dismutase as a Radioprotector". *Radiat. Phys. Chem.* 24, 3/4, 1984, pp. 307–319.

45. Abram Petkau, J.G. Szekely, K.A. Perry: "Simulated responses to intermittant lognormally distributed doses at variable dose rates" *Health Physics, 47,* 5 (November 1984) pp. 745–52.

46. Abram Petkau: "Protection and Repair of Irradiated Membranes" in *Free Radicals, Aging and Degenerative Diseases,* Alan Liss Inc., pp. 481– 508.

47. Abram Petkau : "Scientific basis for the clinical use of superoxide dismutase". *Cancer Treatment Review (198),* 13, pp. 17–44.

48. Abram Petkau: "Protection of bone marrow progenitor cells by superoxide dismutase." *Mol. Cel. Biochemistry,* 1988, 84, pp. 133–40.

49. Jay M. Gould and Benjamin Goldman: *Deadly Deceit: Low-Level Radiation, High-Level Cover-Up* (2nd Edition), 1991, Four Walls Eight Windows, New York.

50. *Personal communication,* June 10, 1990.

51. *EAFV-Bericht,* Vol, 296, October 1987. Birmensdorf, Switzerland.

52. G. Reichelt and R. Kollert: *Waldschäden durch Radioactivität,* 1985, Verlag Müller, Karlsruhe

53. Ralph Graeub: *L'Effet Petkau: Les faibles doses de radioactivité et notre avenir irradié* (1988) Editions d'en bas, Lausanne.

54. G. Reichelt: "Waldschäden in der Umgebung atomtechnischer Anlagen im untered Aaretal bei Beznau/Würenlinger," manuscript, August, 1988.

55. Ralph Graeub: "Apropos Waldschäden im unteren Aaretal," *Natur & Mensch,* Schaffhausen, Switzerland, No. 1, 1989, pp. 18–21.

56. Ralph Graeub: "Apropos Waldschäden im unteren Aaretal," *Basler Zeitung,* September 29, 1988.

57. R. Kollert: "Luftchemie und radioactivität". University of Bremen. *Information Energie und Umwelt,* Part A, No. 27, Nov. 1987. (with Insert Presseinformation, V. 28.6 1988 des Bund "Radioactivität macht luftschadstoffe radikal und sauer").

58. Presseinformation BUND, Freiburg, June 28 1988.

59. C. Hesse: "Der Verdacht." *Magazin des Tagesanzeigers und der Berner Zeitung,* April 14/15, 1949.

60. C. Hesse: *Warum bin ich in Osterfärbebo gewessenℓ,* 1989, Editions Jeuwinkel, Neu Alischwil, Switzerland.

61. A. Masson: "Die heissen Tschernobyl-Teilchen sind in uns." *Berner Zeitung,* October 20, 1986.

62. Jay M. Gould: "Significant U.S. Mortality Increase in the Summer of 1986." Paper presented at the Radiation Victims Conference in New York, September 29, 1987.

63. Jay M. Gould, Ernest J. Sternglass: "Immune System Damage in Cohorts born after the Onset of Nuclear Testing" and "Significant US Mortality Increase after the Chernobyl Accident." Presented at the Symposium on the Effect of Low-level Radiation on Humans, Institute of Radiation Biology, University of Münster, West Germany. February 27, 1988.

64. Ernest J. Sternglass: "Initial Effect of the Fallout from the Chernobyl Accident on the U.S. Population." Paper presented at the Global Radiation Conference, New York, September 29, 1987.

65. Jay M. Gould/Ernest J. Sternglass: "Low-level radiation and mortality" *CHEMTCH,* January 1989, Vol., 19, pp. 18–21.

66. Jens Scheer et al.: "Early Infant Mortality in West Germany before and after Chernobyl." *The Lancet,* November 4, 1989, pp. 1081–2.

67. *ETA-Magazin:* "Tschernobyls American Fallout". June, 1989.

68. *The Lancet:* Letters to the editor. Vol 335, January 20, 1990, pp. 161–2.

69. Kate Millpointer: "Silent Summer." *In These Times,* Vol 13, 4, November 23–December 5, 1988.

70. *The Condor:* 89,1987, pp. 636–53.

71. *The Condor:* 88, 1986, pp. 1–10.

72. *Personal communication,* March 1st, 1990

73. A. Tollmann: Lecture Symp. Anti-Atom-International AAI, 1986 Vienna, (manuscript).

74. John May: *Das Greenpeace Handbuch des Atomzeitalters,* 1989, Droemersche Verlagsanstalt, Münich.

75. Schweiz. Physikal. Gesellschaft, Zürich: "Vom Menschen verursachte Klimaänderundgen." (1989)

76. Richard A. Kerr, *Science:* "Hansen vs. the World on the Greenhouse Threat." June 2, 1989, pp. 1041–43.

77. Richard A. Kerr, *Science:* "How to Fix the Clouds in Greenhouse Models" January 6, 1989, vol 243.

78. *Science:* "Global Warming: Blaming the Sun." Vol. 246, November 24, 1989, pp. 292–3.

79. Herbert Friedmann: *Die Sonne,* Verlag Spektrum der Wissenschaft.

80. *Science*: August 4 1989, Vol 245, pp. 451–2.

81. *Energie und Umwelt:* "Der Treibhauseffekt und die Atomkraft oder wie Argumente sich verflüchtigen" No. 1, March 1988, Energiestiftung, Zürich.

82. R. Kollert and M. Buzin: "Klimagefährdung, Krypton-85." *Globus,* Doppelheft, 10/11, 1988, pp. 324–5.

83. Ralph Graeub: "Apropos Treibhauseffekt-Rebellion von Klimaforschern," *Basler Zeitung,* September 8, 1989.

84. Richard A. Kerr, *Science,* "Greenhouse Skeptic Out in the Cold." December 1, 1989.

85. *PM-Magazin,* "Klima-Katastrophe-ja oder nein?" December, 1989.

86. *Neue Zürcher Zeitung:* "Der Streit um die Klimakatastrophe". January 27/28 1990.

87. *Tages-Anzeiger* (Zürich): "Klimakonferenz in Washington: Bush enttäushcte die Europäer." April 18, 1990.

88. Krause/Bossel: *Energie Wende-Wachstum und Wohlstand ohne Erdöl und Uran,* 1980, S. Fischer Verlag.

89. K.M. Meyer-Abich: *Energie-Energieeinsparung als neue Energiequelle,* 1979, Carl Manser-Verlag.

90. *Schweitzer Energie-Fachbuch 1990.* Künzler-Bachmann AG, St. Gallen, Switzerland.

91. *Energieszenarien* (EGES Report). "Möglichkeiten, Voraussetzungen und Konsequenzen eines Aussteiges der Schweiz aus der Kernenergie." Eidg. Energie-und Verkehrsdep, 1988 edition.

92. B. Ruska and D. Teufel: *Das sanfte Energie-Handbuch,* Rowohlt Taschenbuchveriag, 1982.

93. *ACR-Bulletin:* "BEIR-V Study," American College of Radiology, Vol 46, 1, 1990.

94. John Gofman: *An Irreverent, Illustrated View of Nuclear Power.* San Francisco Committee for Nuclear Responsibility, 1979. pp. 227–8.

95. "The Health Effects of the WIPP Waste Repository." Statement by Dr. Jay Gould, Director RPHP (Letter of June 1989 to the Hearing Officer).

96. *Bundesamt für Gesunheitswesen,* Bern, Switzerland.

97. Ernest J. Sternglass: *Secret Fallout: Low Level Radiation from Hiroshima to Three-Mile-Island,* 1981, McGraw-Hill, New York.

98. J.A. Strand, M.P. Fujihara, T.M. Poston and C.S. Abernathy: "Permanence of Suppression of the Primary Immune Response in Rainbow Trout, Salmo gairdeneri, Sublethally Exposed to Tritiated Water during Emryogenesis." *Radiation Research 91,* 1982, pp. 533–541.

99. M. Morris et al, *South Eastern Massachusetts Health Study* 1978–1990, Division of Environmental Health Assessment, Department of Public Health, Commonwealth of Massachusetts, October 1990.

100. *Greenpeace:* Green, February 1990, p. 22.

101. *Science,* vol. 247, p. 5129, 1990.

Of related interest from Four Walls Eight Windows:
Deadly Deceit: Low-Level Radiation, High-Level Cover-Up
By Dr. Jay M. Gould and Benjamin A. Goldman